ARMY TM 5-600
AIR FORCE AFJPAM 32-1088

BRIDGE INSPECTION, MAINTENANCE, AND REPAIR

JOINT DEPARTMENTS OF THE ARMY AND AIR FORCE
DECEMBER 1994

Published by Books Express Publishing
Copyright © Books Express, 2010
ISBN 978-1-907521-89-8
To purchase copies at discounted prices please contact
info@books-express.com

TECHNICAL MANUAL
No. 5-600
AIR FORCE JOINT PAMPHLET
No. 32-108

HEADQUARTERS
DEPARTMENTS OF THE ARMY
AND THE AIR FORCE
WASHINGTON, DC, *6 December 1994* i

BRIDGE INSPECTION, MAINTENANCE, AND REPAIR

List of Figures

List of Tables

CHAPTER 1

INTRODUCTION

Section I. GENERAL INFORMATION

1-1. Purpose

This manual is a guide for the inspection, maintenance, and repair of bridges for military installations. It is a source of reference for planning, estimating, and technical accomplishment of maintenance and repair work and may serve as a training manual for facilities maintenance personnel in the Army and Air Force engaged in maintenance inspection and repair of bridges.

1-2. Scope

It provides guidance for typical maintenance and

repair of bridges to retain them in continuous readiness for support of military operations. It also describes the methods used in accomplishing this maintenance and repair work. The text includes general principles of maintenance and repair for use by all activities designated to maintain bridges at Army and Air Force installations in a condition suitable for their intended use.

1-3. References

Appendix A contains a list of references used in this manual.

Section II. MAINTENANCE PLANNING

1-4. Programming and economic considerations

In maintenance planning and execution, full consideration must be given to future expected use of each bridge, the life expectancy of the bridge, and the life-cycle cost of periodic repairs versus replacement of a bridge or its major components. The level of maintenance and programming of major repairs should be planned in consonance with future requirement for the bridge and/or planned replacement. The maintenance program should be designed to include prevention of deterioration and damage, prompt detection of deficiencies, and early accomplishment of maintenance and repairs to prevent interruptions of operations or limitation/restriction of bridge use.

1-5. Elements of the maintenance program

a. *Inspection.* Continuous, rigorous inspections are necessary for an effective maintenance program. It is recommended that inspections be made annually of all basic structures and more frequently for fenders and utilities. Additional inspections may be necessary under certain circumstances, such as a tsunami, high tides, earthquakes, accidents, typhoons, and heavy freezes. Inspections may be made from the structures, from a boat or float, or below the waterline by divers. Underwater television is often employed in visual inspections. Types of inspections typical to bridges are:

(1) Operator inspection consists of examination, lubrication, and minor adjustment performed by operators on a continuous basis.

(2) Preventative maintenance inspection is the scheduled examination and minor repair of facilities and systems that would otherwise not be subject to inspection. Pier fender systems, fire protection systems, and under pier utilities are examples.

(3) Control inspection is the major scheduled examination of all components and systems on a periodic basis to determine and document the condition of the bridge and to generate major work required.

b. *Maintenance.* Maintenance is the recurrent day-to-day, periodic, or scheduled work that is required to preserve or restore a bridge to such a condition that it can be effectively utilized for its designed purpose. It includes work undertaken to prevent damage to or deterioration of a bridge that otherwise would be costly to restore. Several levels of bridge maintenance are practiced, depending on the complexity and frequency of the tasks involved. These tasks range from the clearing of drainpipes to the replacement of bearings. Minor maintenance consists of cleaning the drainage system, patch painting, removing debris, tightening loose bolts, and cleaning the joints.

(1) Routine maintenance includes adjusting bearings, complete repainting, repairing potholes, filling cracks, and sealing concrete.

(2) Major maintenance approaches rehabilitation in that it might include the replacement of bearings; readjustment of forces, such as in cables; replacement of joints; fatigue crack repair; waterway adjustment; and other specialized activities not performed very often.

Section III. FREQUENCY OF INSPECTION

1-6. Military requirements

a. Army. AR 420-72 requires that all bridges on all Army installations be thoroughly inspected every year and an analysis of load trying capacities be made every 3 years.

b. Air Force. Bridges on Air Force bases should be inspected at regular intervals not to exceed 2 years. Additionally, bridges should be inspected as soon as possible after severe storms (i.e., floods, hurricanes, etc.) to evaluate possible damage and reduced load-capacity of the structure. A structural analysis of its load-carrying capacity should be performed on at least every third inspection conducted.

1-7. Factors of frequency

a. The depth and frequency to which bridges are inspected will depend on the following factors: age, traffic characteristics, state of maintenance, known deficiencies, and climate conditions.

b. More frequent inspections shall be made if significant change has occurred as a result of floods, excessive loadings, earthquake, or accumulated deterioration. Where changes from those conditions existing in the original analysis and inspection or revalidation are apparent, a reanalysis of the   load-carrying capacity shall be made.

Section IV. QUALIFICATIONS OF INSPECTION PERSONNEL

1-8. Army

a. Annual bridge inspection. Installation maintenance personnel should have a knowledge of bridge structure inspection and construction and be able to identify the regular maintenance requirement and structural deficiency.

b. Triennial bridge inspection. Bridge inspector should have the following minimum qualifications:

(1) Inspector should be a registered professional engineer (or under the direct supervision of a registered professional engineer).

(2) Inspector should have a minimum of 2 years experience in bridge inspection assignment in a responsible capacity.

(3) Inspector should be thoroughly familiar with design and construction features of the bridge to properly interpret what is observed and reported.

(4) Inspector should be capable of determining the safe load carrying capacity of the structure.

(5) Inspector should be able to recognize any structural deficiency, assess its seriousness, and take appropriate action necessary to keep the bridge in a safe condition.

(6) Inspector should also recognize areas of the bridge where a problem is incipient so that preventative maintenance can be properly programmed.

(7) The qualifications of each person directly or indirectly involved with the inspection should be submitted with bid documents.

1-9. Air Force

The Air Force inspector should be a trained bridge inspector from Maintenance Engineering. The main responsibilities are to perform the required inspections, document conditions of the structure, and initiate maintenance actions. If the inspection reveals a situation that requires a greater in-depth evaluation, the inspector (through the division chief) should request a design engineer to evaluate the bridge condition and determine corrective maintenance/repair actions.

CHAPTER 2

BRIDGE STRUCTURES

2-1. Definition

For the purpose of this manual, a bridge is defined as a structure, including supports, erected over a depression or an obstruction, such as water, highway, or railway, having a track or passageway for carrying traffic or other moving loads, and having an opening measured along the center of the roadway of more than 20 feet between undercopings of abutments, or spring lines or arches, or extreme ends of openings for multiple boxes; it may also include multiple pipes, where the clear distance between opening is less than one-half of the smaller contiguous opening.

2-2. Classification

The inspector must be aware of bridge types to properly describe a bridge for the inspection report. The main emphasis of the description should be on the main span. Bridges are classified according to their function, structural type, and structural material:

a. *Function.* The "function" of a bridge refers to the currently approved classification of the roadway. Some typical roadway classifications are: interstate, freeway, principal arterial, minor arterial, collector, major, minor, military, etc.

b. *Structural type.* The "type" of bridge defines both the structural framing system and the type of superstructure:

(1) *Structural framing system.* There are basically four types of structural framing systems: simple spans, continuous spans, cantilever and suspended spans, and rigid frames. They are described as follows:

(a) *Simple span.* These spans consist of a superstructure span having a single unrestrained bearing at each end. The supports must be such that they allow rotation as the span flexes under load. Ordinarily, at least one support is attached in a way that keeps the span from moving longitudinally. Figure 2-1 (part a) demonstrates a simple span.

(b) *Continuous span.* Spans are considered continuous when one continuous piece crosses three or more supports. Figure 2-1 (part b) shows a two-span continuous structure. Note that the supports at the ends of the continuous units are similar to those at the ends of a simple span. However, because the member is continuous over the center support, the magnitude of the member rotation is restricted in the area adjacent to the

pier. A bridge may be continuous over many supports with similar rotational characteristics over each interior support.

(c) *Cantilever and suspended spans.* Sometimes it is advantageous from a structural standpoint to continue a span over the pier and terminate it near the pier with a short cantilever. This cantilever is ordinarily used to support or "suspend" the end of an adjacent span. This arrangement is shown in figure 2-1 (part c). The other end of the suspended span may in some cases be supported by another cantilever or it may rest on an ordinary simple support.

(d) *Rigid frames.* These are frequently used as transverse supports in steel construction and occasionally used as longitudinal spans. The term "rigid" is derived from the manner of construction or fabrication which does not allow relative rotation between the members at a joint. A rigid frame may be rigidly attached at the base (fixed), or it may be simply supported.

(2) *Superstructure type.* The various types of superstructures are: slab, truss, girder, arch, suspension (not covered in this manual), beam-girder, stringer, and composite.

c. *Structural material.* The basic types of structural materials are steel, concrete, timber, stone, masonry, wrought iron, cast iron, and aluminum.

2-3. Typical bridges

Based upon the above classification criteria, many typical bridges can be defined and are summarized in figures 2-2 through 2-5.

2-4. Box culverts

Box culverts range in size from small, single-cell units to multicell units as large as 20 by 20 feet. While natural rock, when present, may be used as a floor, the box culvert is usually a closed, rectangular frame. Usually, transverse joints are provided every 20 to 30 feet. Occasionally, old culverts consist of simply a slab on a wall. These are not true box culverts. Some of these slabs are made of stone, while some walls are made of rubber masonry, rather than concrete.

2-5. Military bridges

A large variety of special-purpose military bridges exist. These bridges, for the most part, are designed for expedient deployment under combat situations. They are not intended for continued day-to-day usage under civilian and military traffic. However, due to economic constraints, some of these bridges are serving as "permanent" structures on some military installations. Therefore, the inspection, maintenance, and repair of these bridges must also be addressed. Several types of military bridges are shown in figures 2-6 through 2-10.

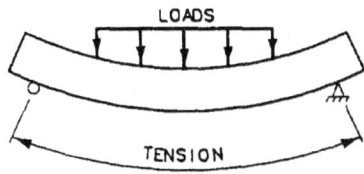

a. Simple beam.

BOLTED SPLICE

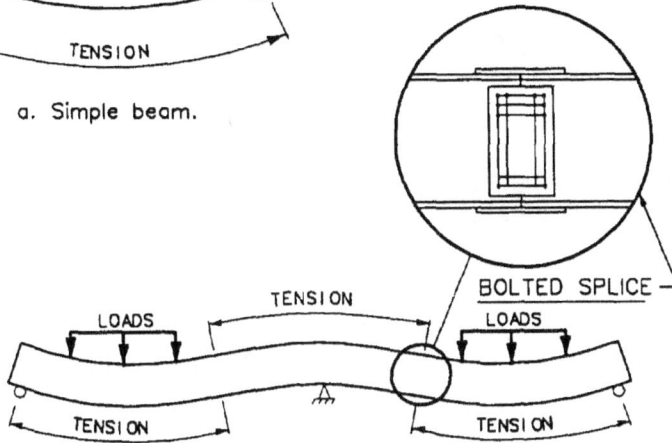

b. Continuous spans.

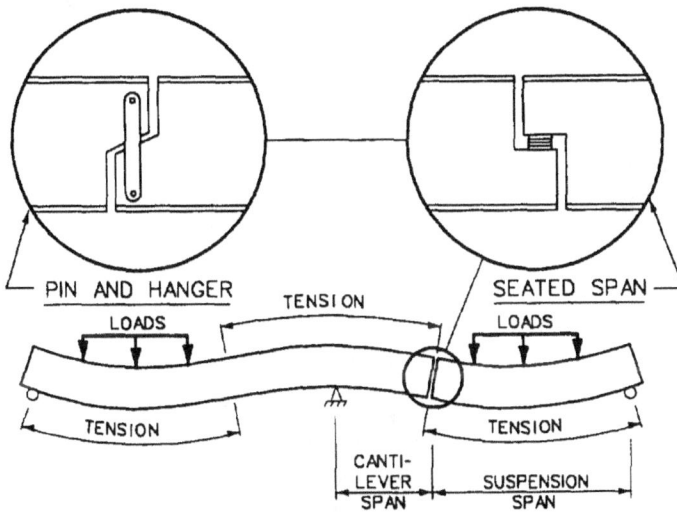

PIN AND HANGER

SEATED SPAN

c. Cantilever-suspended spans.

Figure 2-1. Structural framing system.

THROUGH HOWE TRUSS

THROUGH PRATT TRUSS

THROUGH WARREN TRUSS

QUADRANGULAR THROUGH WARREN TRUSS

THROUGH WHIPPLE TRUSS

CAMEL BACK TRUSS

THROUGH BALTIMORE TRUSS

K-TRUSS

THROUGH TRUSS

PONY TRUSS

DECK TRUSS

CANTILEVER

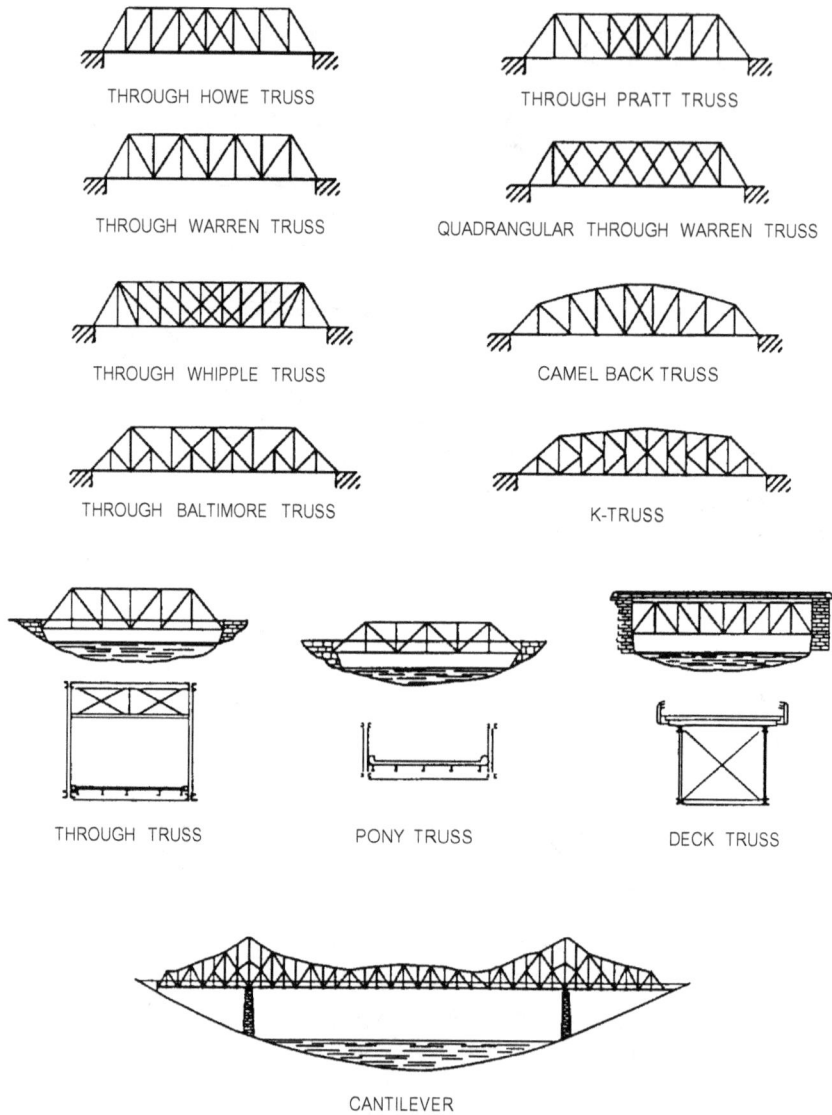

Figure 2-2. Truss bridges: steel or timber construction.

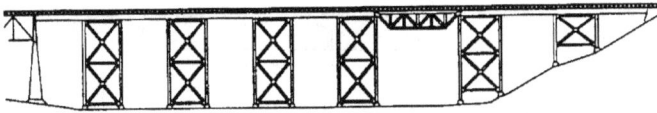

STEEL VIADUCT

THROUGH-ARCH TRUSS

RIGID FRAME-STEEL

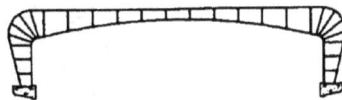

RIGID FRAME
(STEEL GIRDER ELEMENT)

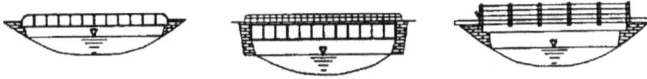

THROUGH GIRDER   DECK GIRDER   1 BEAM

Figure 2-3. Steel bridges.

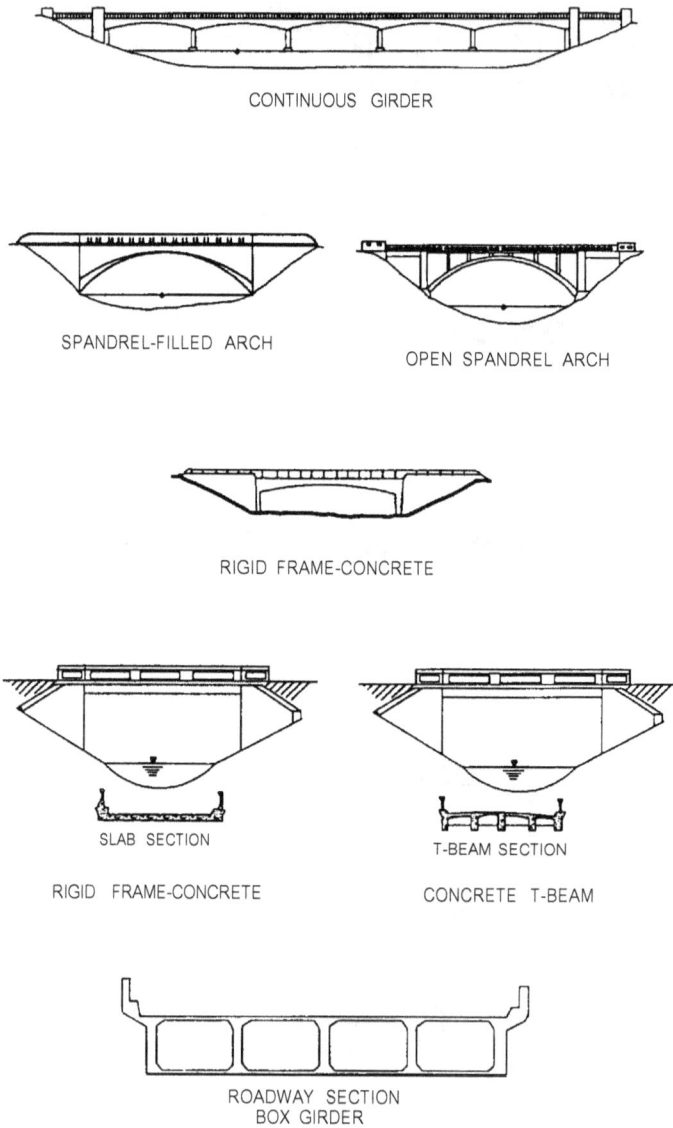

CONTINUOUS GIRDER

SPANDREL-FILLED ARCH

OPEN SPANDREL ARCH

RIGID FRAME-CONCRETE

SLAB SECTION

RIGID FRAME-CONCRETE

T-BEAM SECTION

CONCRETE T-BEAM

ROADWAY SECTION
BOX GIRDER

Figure 2-4. Reinforced concrete bridges.

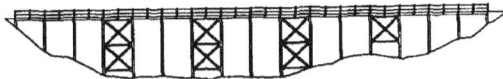

TIMBER TRESTLE

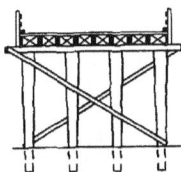 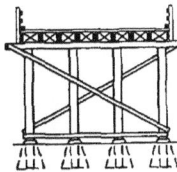

PILE BENT   FRAME BENT

Figure 2-5. Timber bridges.

Figure 2-6. Bailey bridge using bailey panel piers.

Figure 2-7. Double-double bailey bridge.

Figure 2-8. T6 aluminum fixed bridge.

Figure 2-9. Class 50 M-4 trestle bridge, aluminum.

Figure 2-10. Timber trestle.

CHAPTER 3

BRIDGE ELEMENTS

Section I. SUBSTRUCTURE ELEMENTS

3-1. General

Typical bridge nomenclature is summarized in figure 3-1. A bridge basically consists of two main parts: the substructure and the superstructure. The substructure includes those parts which transfer the loads from the bridge span down to the supporting ground. For a single-span structure the substructure consists of two abutments, while for multispan structures there are also one or more piers. Sometimes steel bents or towers are used instead of piers. The loads are applied to the substructure through the bearing plates and transmitted through the abutment walls or pier columns to the footings. If the soil is of adequate strength, the footings will distribute the loads over a sufficiently large area. If not, the footings themselves must be supported on pile foundations extended down to a firm underlying stratum.

3-2. Abutments

Abutments are substructures supporting the end of a single-span or the extreme end of a multispan superstructure and usually retaining or supporting the approach embankment. Typical abutments are shown in figure 3-2. Abutments usually consist of a footing, a stern or breast wall, a bridge seat, a backwall, and wing walls. The backwall prevents the approach embankment soil from spilling onto the bridge seat, where bearings for the superstructure are situated. The wing walls are retainers which keep the embankment soil around the abutment from spilling into the waterway or roadway that is spanned by the bridge. When U-shaped wing walls are used, parapets and railings are often placed on top of them. Abutments may be constructed of plain concrete, reinforced concrete, stone masonry, or a combination of concrete and stone masonry. Plain concrete and stone masonry abutments are usually gravity structures, while reinforced concrete abutments are mostly cantilever or counterfort types.

3-3. Piers and bents

Typical piers and bents are shown in figure 3-3. Piers transmit the load of the superstructure to the foundation material and provide intermediate supports between the abutments. Footings, columns or stems, and caps are the main elements of piers. The footings are slabs which transmit the load to the soil, rock, or to some other foundation unit such as piles, caissons, or drilled shafts. The columns or stems transmit vertical load and moment to the footings. The cap receives and distributes the superstructure loads. River bridges, railway bridges, and some highway underpasses are likely to use the solid wall pier. Highway grade separations of normal width often use multilegged piers, often with a cap binding the whole unit into a rigid frame. "Bents" are basically piers without footings, which consist of a row of two or more posts or piles, tied together at the top with a cap. Piers and bents may be made of timber, steel, concrete, stones, or combination of materials. Piles are used to transmit the bridge loads to the foundation material when the foundations are to be on soft soils, in deep water, or in swift streams. Typical pile types are: steel H Piles, timber, concrete piles (both CIP and precast/prestressed), and concrete filled pipe or shell piles.

Section II. SUPERSTRUCTURES

3-4. General

The superstructure includes all those parts which are supported by the substructure, with the main part being the bridge spans. Vehicular forces are transmitted from the bridge deck, through the supporting beams or girders of the span, and into the substructure. The reinforced concrete slab bridge has the simplest type of superstructure since the slab carries the load of the vehicle directly to the abutment or piers. On beam or girder bridges, the slab is supported on longitudinal steel, concrete, or timber members which, in turn, carry the load to the abutment or piers. Some superstructures consist of the deck, a floor system, and two or more main supporting members. Figure 3-4 shows several different types of superstructures and their associated elements. The components of superstructures are summarized in the following paragraphs.

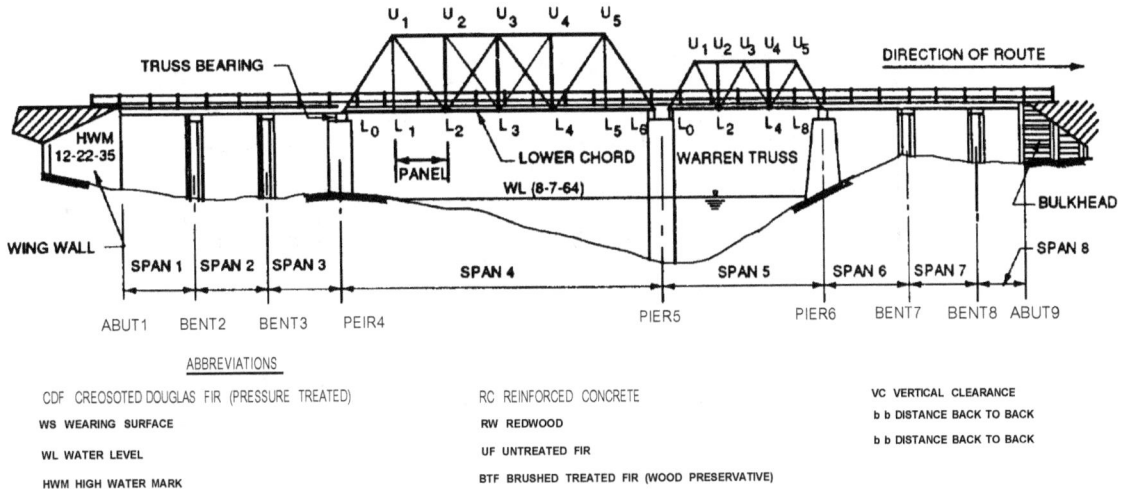

Figure 3-1. Bridge nomenclature.

3-5. Decks

The deck is that portion of a bridge which provides direct support for vehicular and pedestrian traffic. The deck may be a reinforced concrete slab, timber flooring, or a steel plate or grating on the top surface of abutting concrete members or units. While normally distributing load to a system of beams and stringers, a deck may also be the main supporting element of a bridge, as with a reinforced concrete slab structure or a laminated bridge. The "wearing course" of a deck provides the riding surface for the traffic and is placed on top of the structural portion of the deck. There are also wearing courses poured integral with the structural slab, and then the deck is referred to as a monolithic deck.

3-6. Floor systems

The floor system may consist of closely spaced transverse floor beams between girders (refer to the deck girder bridge in figure 3-4, sheet 2, part c) or several longitudinal stringers carried by transverse floor beams (refer to the through girder of figure 3-4, sheet 2, part c). In floors of this type, the stringers are usually wide flange beams while the floor beams may be plate girders, wide flange beams, or trusses. Where floor beams only are used, they may be rolled beams or plate girders. Several floor systems are shown in figure 3-5.

3-7. Main supporting members

The main supporting members transmit all loads from the floor system to the supports at points on the piers and abutments. The strength and safety of the bridge structure depends primarily on the main supporting members. These members may be timber, steel, or concrete beams; steel plate girders; timber or steel trusses or concrete rigid frames; arches of various material; or steel cables. The most general types of these members are discussed as follows:

a. Rolled beams. The rolled beam is used for short spans. It comes from the rolling mill as an integral unit composed of two flanges and a web. The flanges resist the bending moment and the web resists shear. The more common types of rolled beam shapes are shown in figure 3-6.

b. Plate (built-up) girders. This type of structural member is used for intermediate span lengths not requiring a truss and yet requiring a member larger than a rolled beam. The basic elements of a plate girder are a web to which flanges are riveted or welded at the top and bottom edges. The most common forms of cross section are shown in figure 3-7. The component parts of a plate girder are as follows (figure 3-7):

(1) *Flange angles.* These are used for riveted plate girders and carry tensile or compressive forces induced by bending.

FULL HEIGHT ABUTMENT

STUB ABUTMENT

OPEN ABUTMENT

TYPICAL CONCRETE ABUTMENTS

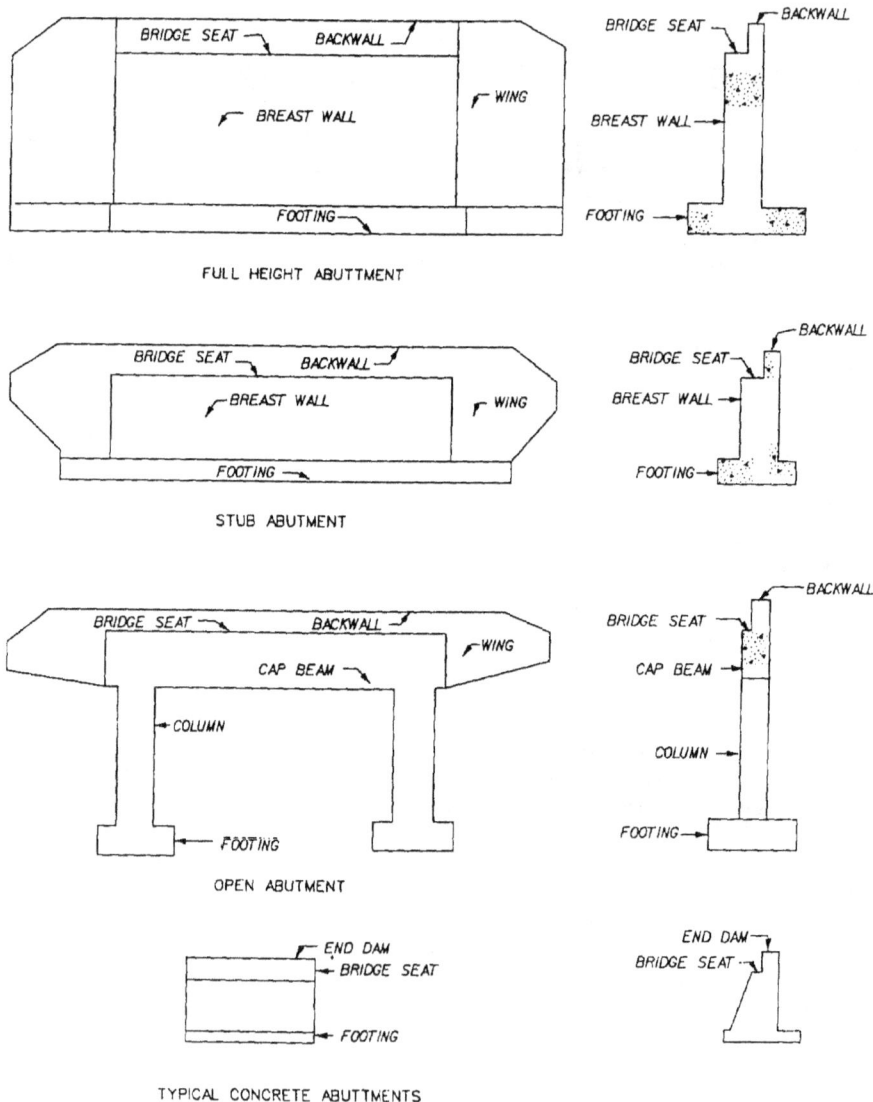

Figure 3-2. Typical abutments

(2) *Cover plates.* These are welded or riveted to the top and/or bottom flanges of the girder to increase the load carrying capacity.

(3) *Bearing stiffeners.* These are plates or angles placed vertically at the locations of the support and attached to the web. Their primary function is to transmit the shearing stresses in the web plate to the bearing device and, by so doing, prevent web crippling and buckling.

(4) *Intermediate stiffeners.* These are used at points of concentrated loads or for deep girder to prevent web crippling and buckling.

c. Concrete beams. These beams are usually reinforced wherein the tensile stresses, whether resulting from bending, shear, or combinations thereof produced by live and dead loadings, are by design carried by the metal reinforcement. The concrete takes compression (and some shear) only. It is commonly rectangular or tee-shaped with its depth dimension greater than its stem width.

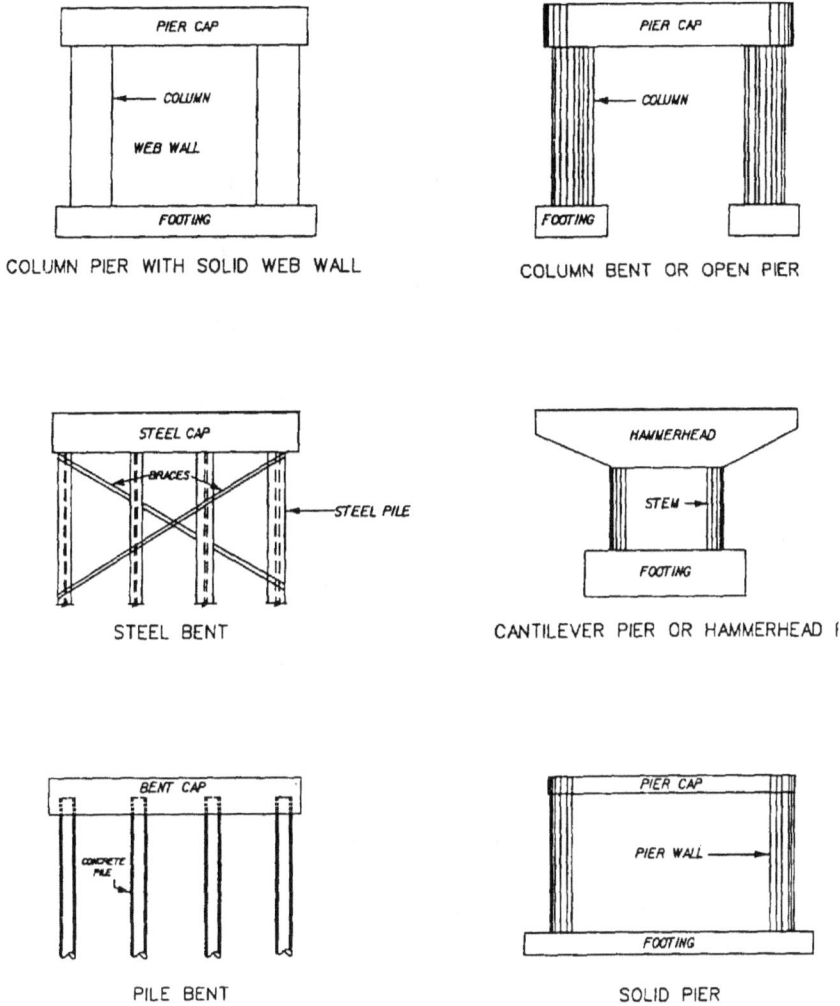

Figure 3-3. Typical piers and bents.

d. Trusses. The truss is one form of structural system which, because of its characteristics, provides high load-carrying capacities and can be used to span greater lengths than rolled beams and girders. The truss functions basically in the same manner as a rolled beam or girder in resisting loads, with the top and bottom chords acting as the flange and the beam and the diagonal members acting as the web. While most trusses are of steel, timber trusses also exist. Truss members may be connected with rivets, bolts, or pins. Although the configuration of trusses varies widely, the essential components are common to all. Truss members may be built-up sections, rolled sections, tubing, pipe, eyebars, or solid rods. An earlier commonly used construction practice was to connect channels by lacing bars and stay plates at the ends. Interior verticals and diagonals on old bridges may consist of relatively slender solid rods when the member is subject only to tension. When two opposing tension diagonals are provided in the panel of a truss, they are termed "counters." The basic parts of a truss are summarized in figures 3-1, 3-4 (sheet 1, part b) and 3-8. They are discussed as follows:

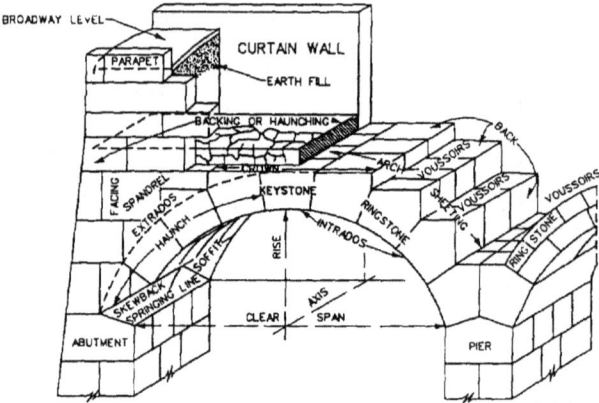

A. MASONRY ARCH BRIDGE.

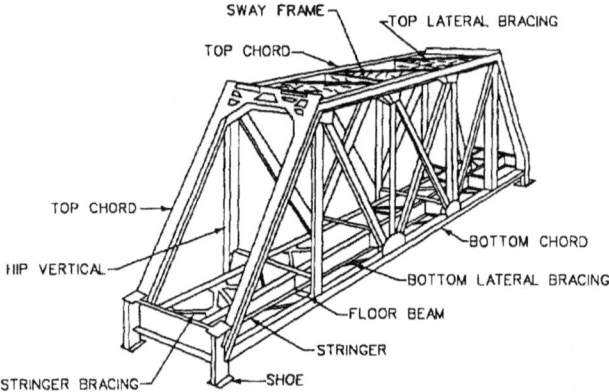

B. TRUSS

Figure 3-4. Typical superstructures. (Sheet 1 of 2)

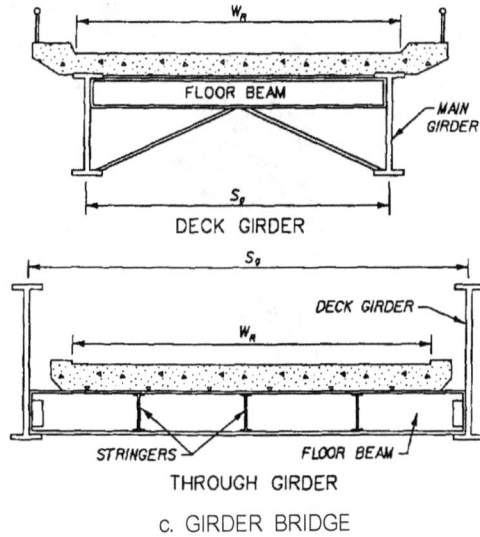

c. GIRDER BRIDGE

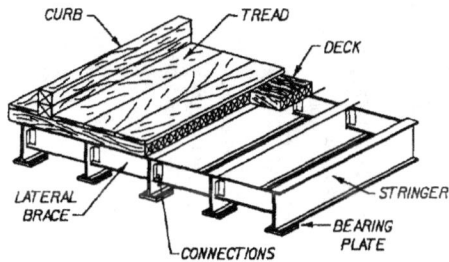

d. STEEL STRINGER BRIDGE

Figure 3-4. Typical superstructures. (Sheet 2 of 2)

(1) *Chord.* In a truss, the upper and the lower longitudinal members extending the full span length are called chords. The upper portion is designated as the upper or top chord and correspondingly lower portion is designated as the lower or bottom chord. For simple span, the top chord will always be in compression, and the bottom chord will always be in tension and should be considered a main structural member. Failure of either chord will render the truss unsafe.

(2) *Diagonals.* The diagonal web members span between successive top and bottom chords and will resist tension or compression, depending on the truss configuration. Most diagonals are also main structural members and their failure would be extremely critical and render the truss unsafe.

(3) *Verticals.* Vertical web members span between top and bottom chords, which will resist tension or compression stresses depending upon the truss configuration. Most verticals are a main structural member, and their failure would usually be critical and render the truss unsafe.

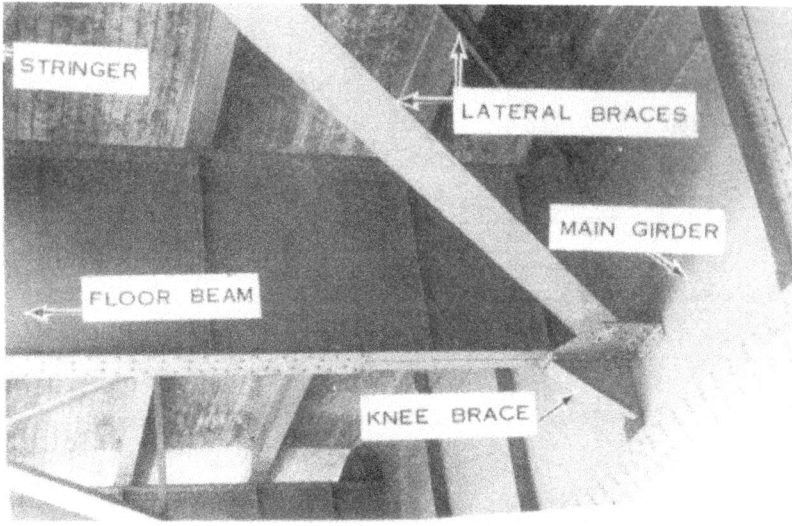

Figure 3-5. Typical floor systems. (Sheet 1 of 3)

Figure 3-5. Typical floor systems. (Sheet 2 of 3)

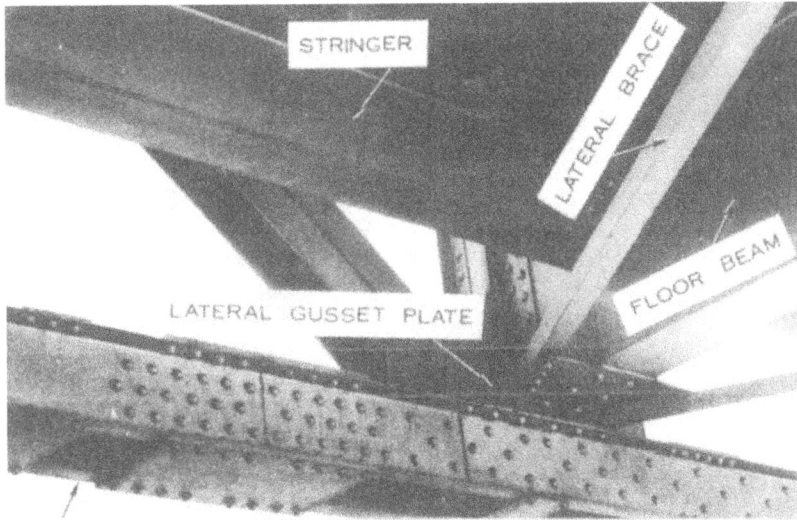

Figure 3-5. Typical floor systems. (Sheet 3 of 3)

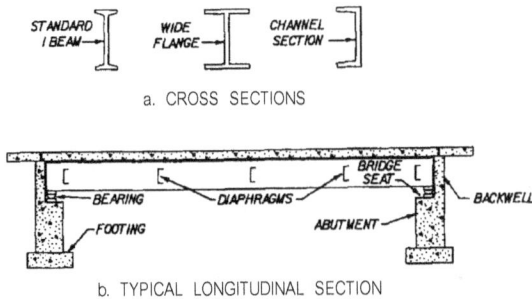

Figure 3-6. Rolled steel beams.

(4) *Panel point.* The panel point is the point of intersection of primary web and chord members of a truss. Note: Items 5 through 11 can be considered secondary structural members and although their failure should receive immediate attention, an individual member failure will not render the structure unsafe.

(5) *Portal bracing.* The portal bracing is found overhead at the end of a through truss and provides lateral stability and shear transfer between trusses.

(6) *Sway bracing.* This bracing spans between the trusses at interior panel points and provides lateral stability and shear transfer between trusses.

(7) *Top lateral bracing.* The top lateral braces lie in the plane of the top chord and provide lateral stability between the two trusses and resistance to wing stress.

(8) *Bottom lateral bracing.* These braces lie in the plane of the bottom chord and provide lateral stability and resistance to wind stresses.

(9) *Floor beam.* The floor beam spans between trusses at the panel points and carries loads from the floor stringer and deck system to the trusses.

(10) *Stringers.* These span between floor beams and provide the primary support for the deck system. The deck loading is transmitted to the stringers and through the stringers to the floor beams and to the truss.

(11) *Gusset plates.* These plates connect the structural members of a truss. On older trusses, pins are used instead of gussets.

3-8. Bracing

The individual members of beam and girder structures are tied together with diaphragms and cross frames; trusses are tied together with portals, cross frames, and sway bracing. Diaphragms and cross frames stabilize the beams or trusses and distribute loads between them (figure 3-9). A diaphragm is usually a solid web member, either of a rolled shape or built up, while a cross frame is a truss, panel, or frame. Since portals and sway braces help maintain the cross section of the bridge, they are positioned as deep as clearance requirements permit. Portals usually are in the plane of the end posts and carry lateral forces from the top chord bracing to the supports (figure 3-8, sheet 1). Lateral bracing placed at the upper or lower chords (or flanges), or at both levels, transmits lateral forces (such as wind) to the supports (figures 3-8 (sheet 3), 3-8 (sheet 4), and 3-9).

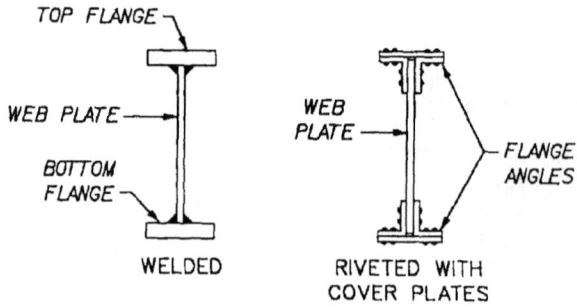

a. CROSS SECTIONS

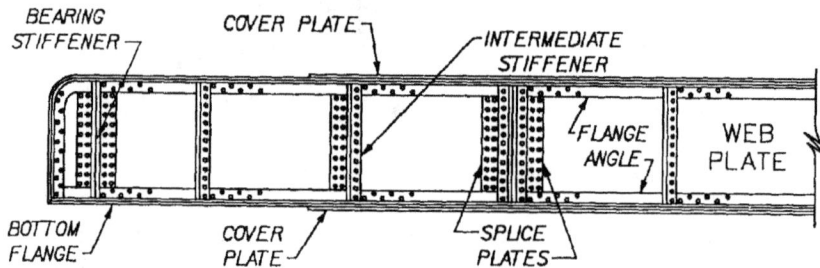

b. TYPICAL LONGITUDINAL SECTION

Figure 3-7. Plate girders.

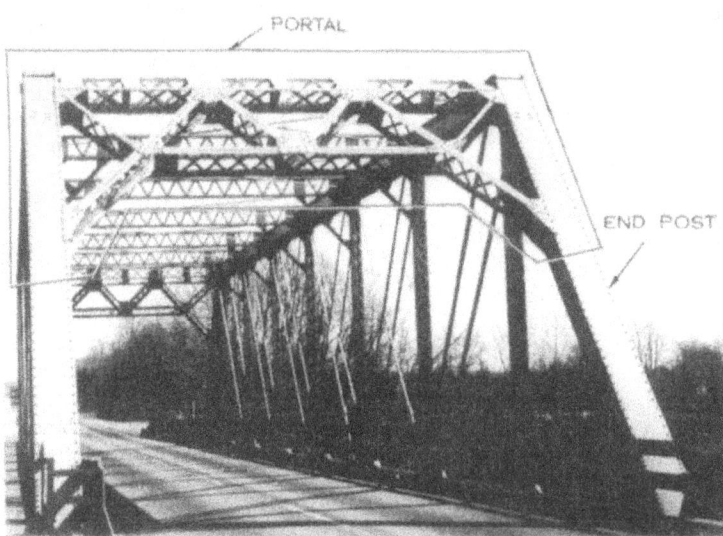

Figure 3-8. Truss components. (Sheet 1 of 5)

Figure 3-8. Truss components. (Sheet 2 of 5)

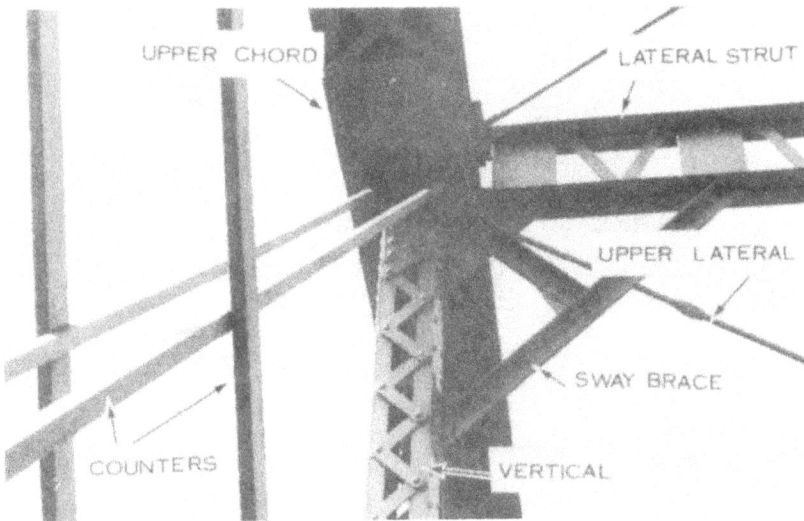

Figure 3-8. Truss components. (Sheet 3 of 5)

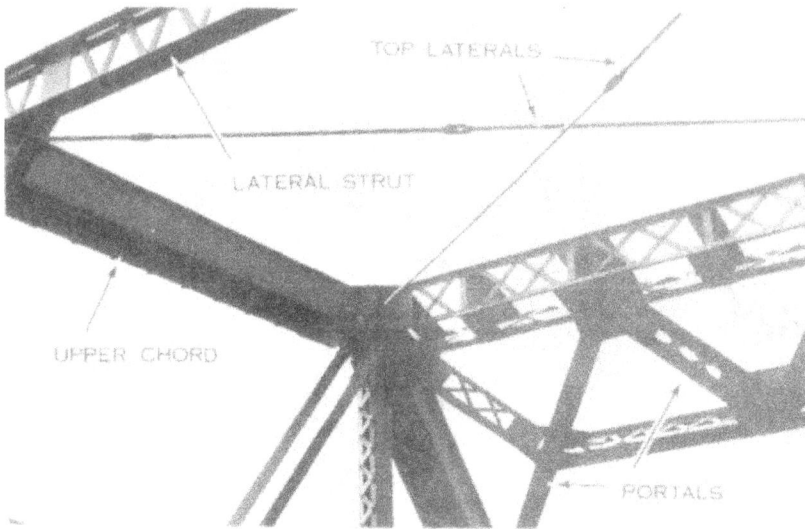

Figure 3-8. Truss components. Sheet 4 of 5)

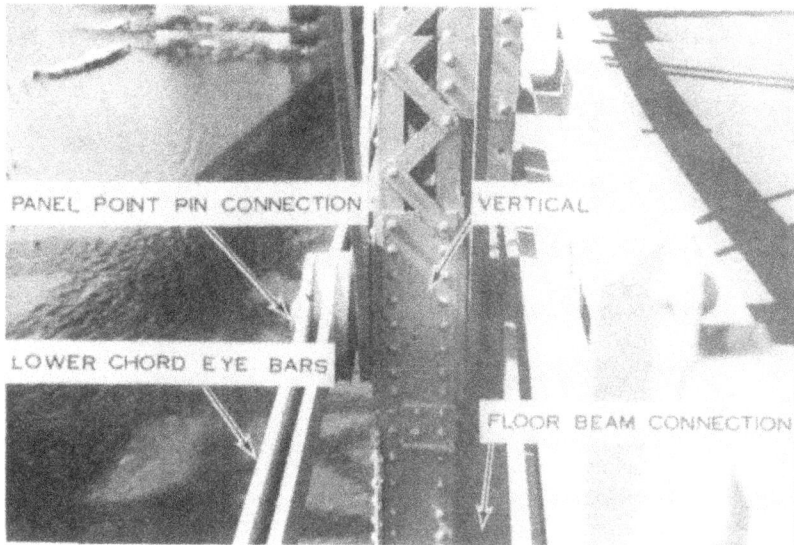

Figure 3-8. Truss components. (Sheet 5 of 5)

Figure 3-9. Bracing. (Sheet 1 of 2)

Figure 3-9. Bracing. (Sheet 2 of 2)

Section III. MISCELLANEOUS ELEMENTS

3-9. Bearings

Bearings transmit and distribute the superstructure loads to the substructure, and they permit the superstructure to undergo necessary movements without developing harmful overstresses. They are vitally important to the functioning of the structure. If they are not kept in good working order, very high stresses may be induced in the structure that could shorten the usable life of the structure. Bearings are of two general types, fixed and expansion. Fixed bearings resist lateral and longitudinal movement of the superstructure but permit rotation. Expansion bearings allow longitudinal movement to account for expansion and contraction of the superstructure. Depending on structural requirements, the bearings may or may not be designed to resist vertical uplift. Bearings are metal or elastomeric. Typical metal bearings are shown in figures 3-10 and 3-11. Elastomeric bearing pads (figure 3-12) have become a popular choice for use as expansion bearings. They are made of a rubber-like material or elastomer molded in rectangular pads, or in strips. Note that bearings can also be used to support suspended spans as shown in figure 3-13.

3-10. Pin and hanger supports

These are devices used to attach a suspended section to a cantilevered section (figure 3-14). These connections may be free or fixed at one end as shown in figures 3-15 and 3-16.

3-11. Expansion joints

Since all materials expand and contract with changes in temperature, provisions must be made in the bridge superstructure to permit movement to take place without damage to the bridge. On very short superstructures, there is usually sufficient yielding in the foundation to allow the small amount of movement to occur without difficulty. On longer structures, however, specifically designed expansion joints are provided in the deck. Where only moderate amounts of movements are expected, the joint may be only an opening between abutting parts. When a watertight seal is desired, a premolded filler topped with a poured-in-place sealer or a preformed compression seal is inserted (figure 3-17, sheet 1). Where traffic is heavy, the unprotected edge of the joint is usually armored with steel angels set in the concrete. When larger movements must be accommodated, a sliding plate or finger plate expansion joint may be used (figure 3-17, sheet 2). A trough is often provided beneath a finger plate expansion joint to catch water from the roadway.

3-12. Approaches

The approach provides a smooth transition between the roadway pavement and the bridge deck. This is important because it reduces impact forces acting on the bridge. Rough approaches are usually the result of a volume change either from settlement in the backfill material or from a general consolidation of subsoil and approach fills, while the bridge, supported on piles, does not settle at all. To avoid problems from differential settlement, approach slabs are often used which span the 15 to 25 feet of fill immediately behind the abutments.

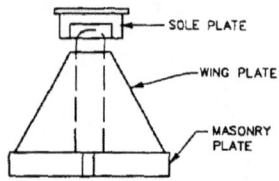

A. FIXED BEARING.

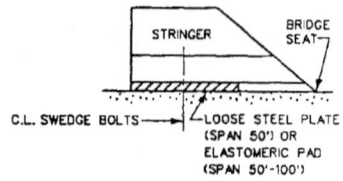

B. SIMPLE EXPANSION BEARING

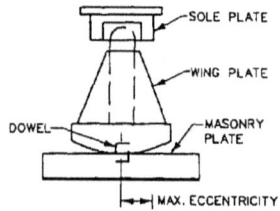

C. EXPANSION BEARING.

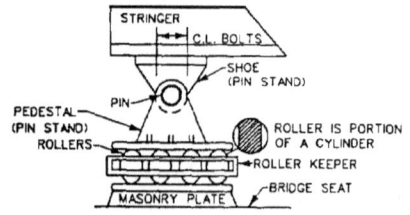

D. ROLLER EXPANSION BEARING

Figure 3-10. Metal bearing types.

Figure 3-11. Metal bearings. (Sheet 1 of 4)

Figure 3-11. Metal bearings. (Sheet 3 of 4)

Figure 3-11. Metal bearings. (Sheet 2 of 4)

Figure 3-11. Metal bearings. (Sheet 4 of 4)

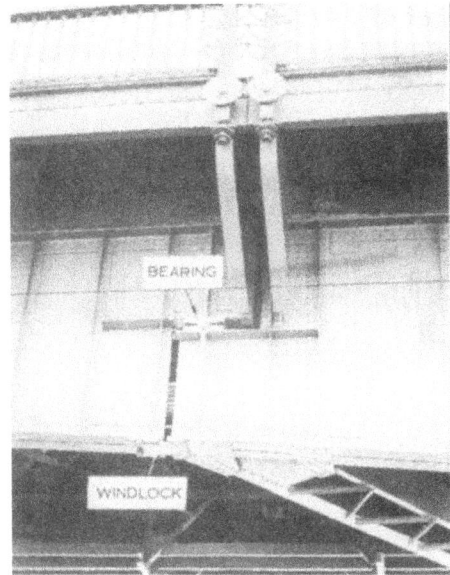

Figure 3-12. Elastomeric bearings.

Figure 3-13. Bearing support for a suspended span.

3-13. Railings, sidewalks, and curbs

a. *Railings.* Railings should be sufficiently strong to prevent an out-of-control vehicle from going off the bridge. However, many existing bridges have vehicle guard rails that are little better than pedestrian handrails. Such guard rails are inadequate for safety, are easily damaged by vehicles, and are susceptible to deterioration. On the other hand, an unyielding guard rail poses a hazard to vehicular traffic particularly if struck head on. Unprotected parapets pose a similar danger.

b. *Sidewalks.* Sidewalks are provided to protect pedestrians crossing the bridge.

c. *Curbs.* A curb is a stone, concrete, or wooden barrier paralleling the side unit of the roadway to guide the movement of vehicle wheels and safeguard bridge trusses, railings, or other constructions existing outside of the roadway unit and also pedestrian traffic upon sidewalks from collision with vehicles and their loads.

3-14. Deck drains

Bridge drainage is very important since trapped or ponded water, especially in colder climates, can cause a great deal of damage to a bridge and is a safety hazard. Therefore, an effective system of drainage that carries the water away as quickly as possible is essential to the proper maintenance of the bridge.

3-15. Utilities

It is common for commercial and industrial utilities to use highway rights-of-way and/or adjacent areas to provide goods and services to the public. This means that some utility operations will be found on a number of bridge structures. These operations may be one or more of the following: gas, electricity, water, telephone, sewage, and liquid fuels. Utility companies perform most of their

Figure 3-14. Pin and hanger connection.

facilities installation and most of the required maintenance. While the large commercial enterprises, e.g., gas, light, and telephone companies, will usually perform the scheduled maintenance of their facilities, some of the smaller publicly owned utilities, e.g., water companies, are less likely to perform adequate maintenance since they may not be as well staffed. Most utility lines or pipes are suspended from bridges between the beams or behind the fascia. On older bridges, water pipes and sewer pipes may be installed along the sides of the bridge or may be suspended under the bridge.

3-16. Lighting

Lighting on bridges will consist of "whiteway" lighting, sign lights, traffic control lights, navigation lights, and aerial obstruction lights. The last two types of lights are special categories which are encountered only on bridges over navigable waterways or on bridges having high towers. There will, of course, be many bridges with no lighting at all.

3-17. Dolphins and fenders

Dolphins and fenders around bridges protect the structure against collision by maneuvering vessels. The fender system absorbs the energy of physical contact with the vessel. The various types of dolphins and fenders are as follows:

  a. *Dolphins.*
  (1) *Timber pile clusters.* This type of dolphin is widely used and consists of a cluster of timber piles driven into the harbor bottom with the tops pulled together and wrapped tightly with wire rope (figure 3-18).

  (2) *Steel tubes.* Steel tube dolphins are composed of one or more steel tubes driven into the harbor bottom and connected at the top with bracing and fendering systems.

  (3) Caissons. These are sand-filled, sheet-pile cylinders of large diameter. The top is covered by a concrete slab, and fendering is attached to the outside of the sheets.

  b. *Fenders.*
  (1) *Timber bents.* A series of timber piles with timber walers and braces attached to the tops are still used (figure 3-19). Steel piles are sometimes used in lieu of timber.

  (2) *Cofferdams.* On large bridges with wide footings, the cofferdam sheets left in place and braced by a concrete wall act as pier protection. A grid or grillage of timber or other resilient material on the outside of the sheets forms a collision mat.

  (3) *Steel pile fenders.* Steel piles driven in pairs to form a frame, with a concrete slab tying the piles together, make a good fender. Timber grillages are attached to the outside to absorb collision impact.

  (4) *Steel or concrete frames.* Steel or concrete frames are sometimes cantilevered from the pier and faced with a timber or rubber cushioning to reduce collision impact.

  (5) *Timber grids.* Timber grids, consisting of posts and walers, are attached directly to the pier (figure 3-18).

  (6) *Floating fenders.* Floating frameworks which partly or completely surround the pier are sometimes used as fenders. The main frames are

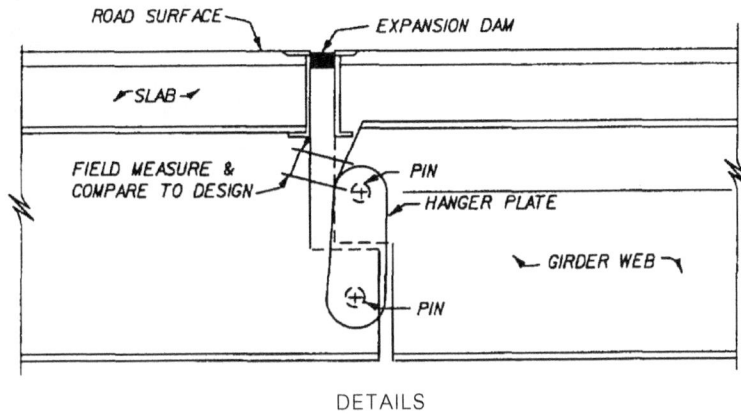

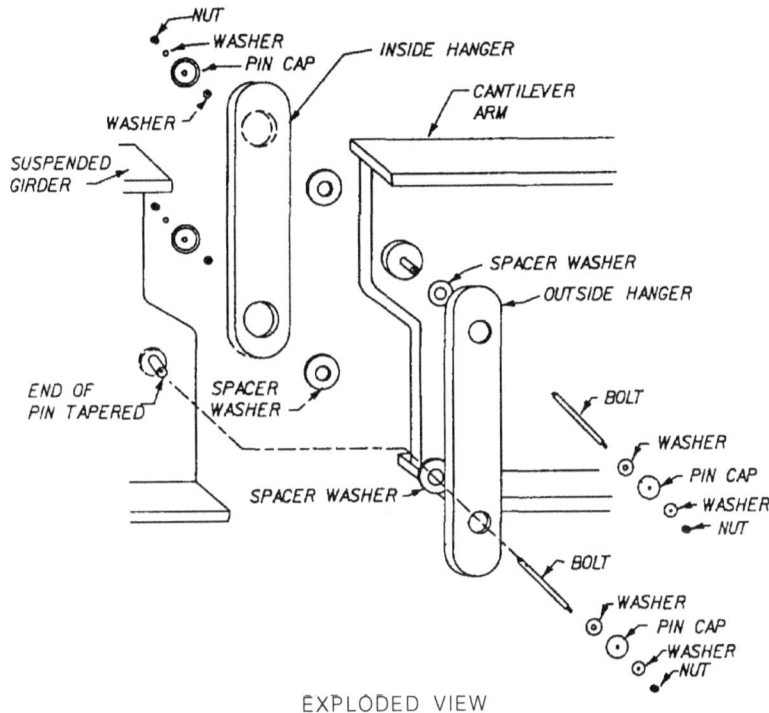

Figure 3-15. Free pin and hanger connection.

usually made of steel or concrete with timber cushioning on the outside face.

(7) *Butyl rubber fenders.* Butyl rubber may be used as a fendering system.

3-18. Welds, bolts, and rivets

a. Welds. Welding is a method of joining two metals together by melting metal at the joints and fusing it with an additional metal from a welding rod. When cool, weld metal and base metal form a continuous and almost homogeneous joint. The two basic types of welds are shown in figure 3-20.

b. Bolts. The A325 high-strength bolt has become the primary field fastener of structural steel. Specifications usually call for a heavy hexagon structural bolt, a heavy semifinished hexagon nut, and one or two washers. Bevel washers may be required.

c. Rivets. Rivets are sometimes used instead of bolts, especially in older structures.

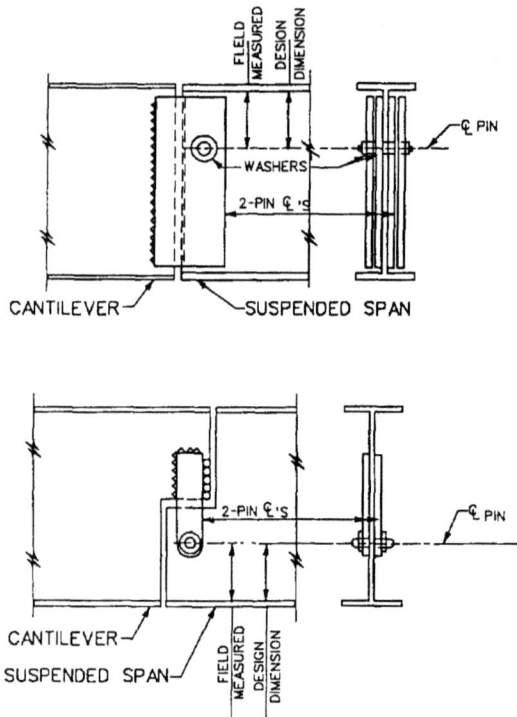

Figure 3-16. Fixed pin and hanger connection.

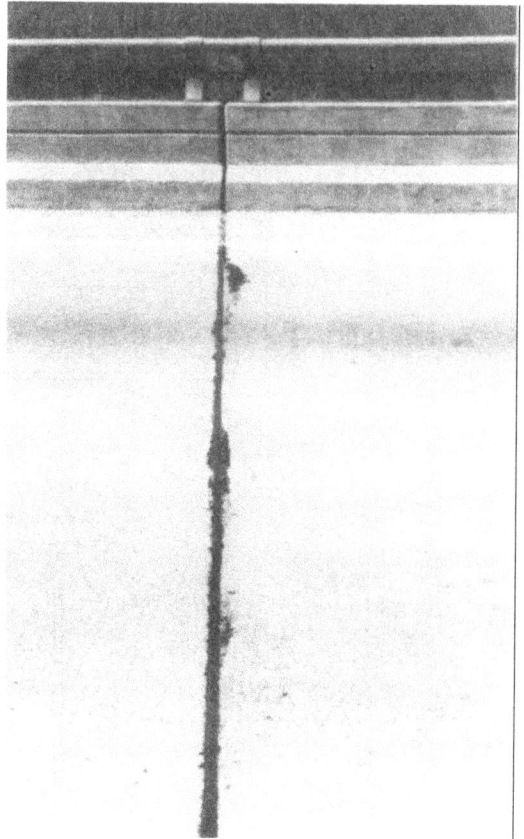

Figure 3-17. Expansion joints. (Sheet 1 of 2)

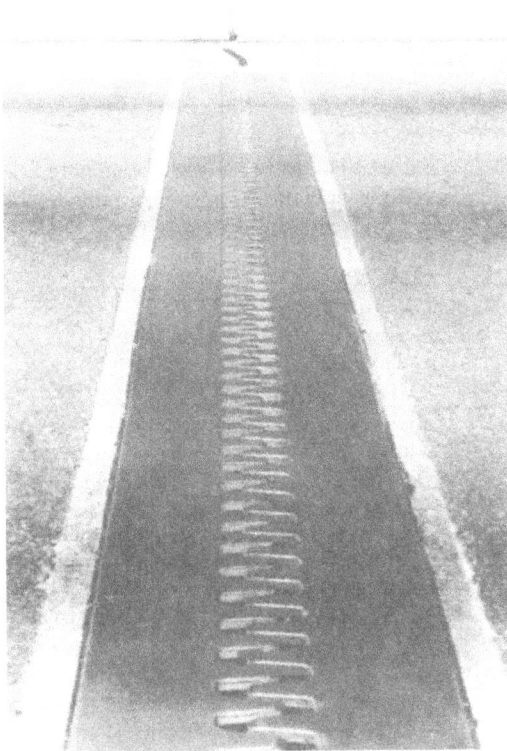

Figure 3-17. Expansion joints. (Sheet 2 of 2)

Figure 3-18. Timber pile cluster dolphins and timber fenders.

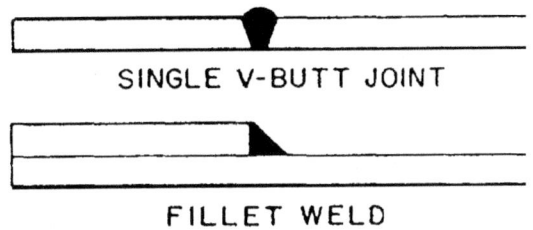

Figure 3-19. Timber bent fenders.

SINGLE V-BUTT JOINT

FILLET WELD

Figure 3-20. Basic weld types.

CHAPTER 4

MECHANICS OF BRIDGES

4-1. General

This chapter provides the inspector with the basics of bridge structures. A more thorough understanding of the behavior of bridges, and the forces imposed on them, will help the inspector to better understand the importance of his job and its associated critical aspects.

4-2. Bridge forces

The solid bodies to be considered are the substructure and the superstructure of the bridges, while the forces exerted on them (represented by arrows in the following diagrams) include various combinations of loads. The principal force is that of gravity acting on the weight of the structure itself (figure 4-1), on the vehicles (figure 4-2), or on other live loads being carried by the structure. Other forces to be considered are those created by earth pressures (figure 4-3); buoyancy or uplift on that portion of a structure which is submerged below water level (figure 4-4); wind loads on the structure, vehicles, or live loads (figure 4-5); longitudinal forces due to changes in speed of the live load or due to friction at the expansion bearings (figure 4-6); temperature change (figure 4-7), earthquake (figure 4-8) stream flow and ice pressure (figure 4-9); and, in the case of masonry structures, shrinkage and elastic rib shortening.

4-3. Stress

The load per unit of area is called unit stress. Unit stress is a very widely used standard for measurement of safe load. Generally, a limiting unit stress is established for a given material. This allowable unit stress multiplied by the cross-sectional area gives the safe load for the member. Since this manual can give only a very elementary introduction to the mechanics of structures, it will be limited to a consideration of the forces due to dead and live loads acting on simple tension or compression members of simple-span structures. For an understanding of other forces and other types of structures, it is suggested that the bridge inspector refer to standard structural analysis textbooks. Loads or forces acting upon members may be classified as axial, transverse, rotational, and torsional. Figures 4-10 and 4-11 illustrate the action of these forces. Both axial and transverse forces are gravity forces and are expressed in pounds, kips (1,000 pounds), tons, or kilograms. When the axial or longitudinal loads exert a pull on the

member, the force is said to be tensile; when the axial load pushes or squeezes a member, the force is compressive. In the pure case, axial forces load the cross-sectional area uniformly as shown in figure 4-11. The formula for axial stress is:

$$f = P \, / \, A \qquad\qquad \text{(eq 4-1)}$$

where

f = stress

P = load

A = cross-sectional area

a. Tension. A simple tension member could be one of the subvertical members of a through truss (figure 4-12). Both dead and live loads cause downward vertical forces which pass from the roadway slab through the stringers and floor beams, each adding its own dead weight force to that already being exerted on it. These combined forces are applied to the subvertical member in question through the floor beam connection to the truss. The tensile force acts on the entire cross section (less rivet or other holes) of the member and produces a certain intensity of stress. If that intensity, or unit stress, is within allowable limits, the member can withstand the applied loads and the member can be considered "safe." If, however, corrosion has reduced the effective area of the member, the intensity of the stress is increased and may exceed the allowable limit. Corrosion may also cause a notch effect which concentrates the stress and further weakens the member.

b. Compression. A simple compression member could be a vertical steel column of a viaduct (figure 4-13). Here the dead and live loads cause downward forces which produce a certain intensity of compressive stress on the entire cross section of the member. In compression members the unit stress not only has to be within allowable limits, as is the case with tension members, but the allowable stress becomes smaller as the slenderness ratio becomes greater. That is, for any given cross section, the longer the column the lower the allowable stress in compression. This is because long compression members will buckle rather than crush.

c. Shear. Transverse forces exert a shearing force or tendency to slide the part of a member to one side of a cross section transversely with respect to the part of the member on the other side of the section. This scissor-like action is illustrated in figure 4-14. Oddly enough, the real shear stress produced by a transverse load is manifested in a

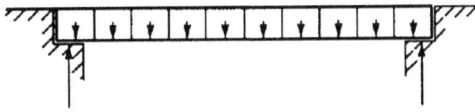

Figure 4-1. Dead load on simple span.

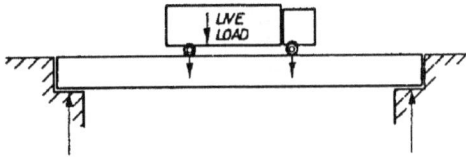

Figure 4-2. Live load on simple span.

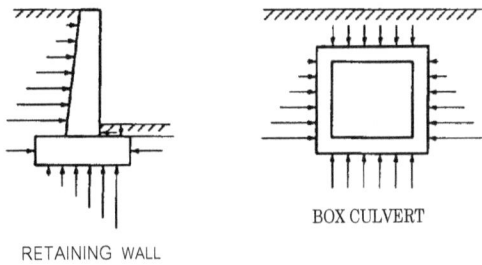

RETAINING WALL

BOX CULVERT

Figure 4-3. Earth pressure.

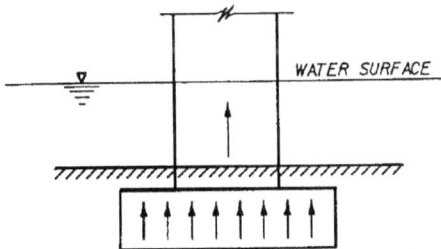

Figure 4-4. Buoyancy on pier.

Figure 4-5. Lateral wind load (end view).

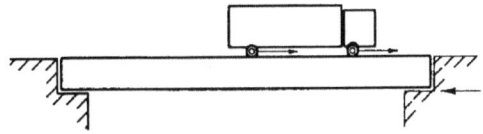

Figure 4-6. Longitudinal force due to friction and live load.

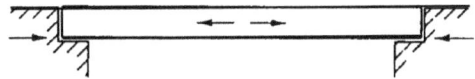

Figure 4-7. Forces due to temperature rise.

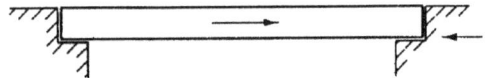

Figure 4-8. Earthquake forces (may be in any direction).

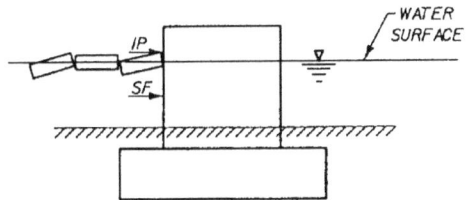

Figure 4-9. Ice pressure and stream flow against pier.

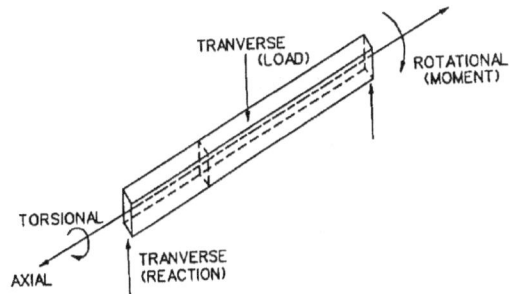

Figure 4-10. Forces on a member.

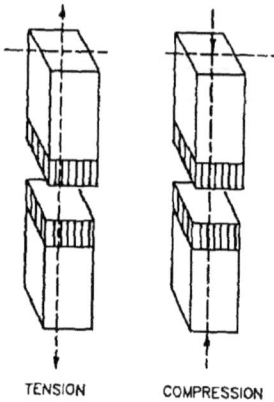

TENSION   COMPRESSION

Figure 4-11. Axial forces and stress.

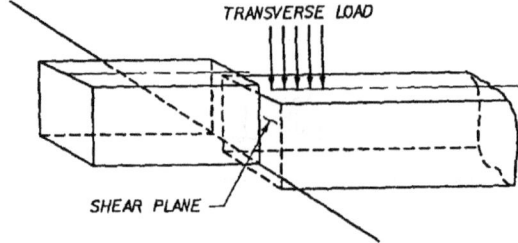

Figure 4-14. Shear forces.

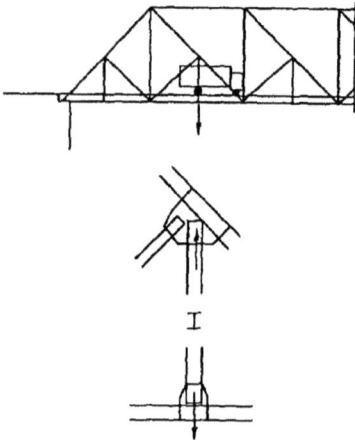

Figure 4-12. Truss vertical in tension.

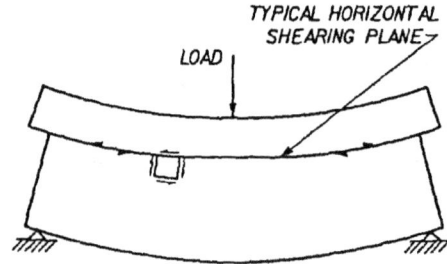

Figure 4-15. Shear. (Sheet 1 of 2)

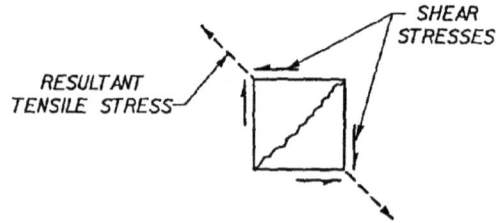

Figure 4-15. Shear. (Sheet 2 of 2)

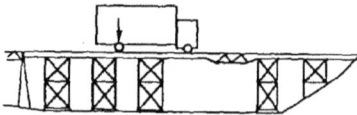

Figure 4-13. Compression.

horizontal shear stress (figure 4-15, sheet 1). However, it is accompanied by a vertical shear stress of equal magnitude as shown in figure 4-15 (sheet 2), which is an enlargement of the little element of figure 4-15 (sheet 1). It can easily be seen that the four shear stresses will combine to form a tensile stress. While this is the most likely source of shear problems, most design criteria consider vertical shear as the criterion of shear strength. The formula for vertical shear stress is:

$$f_v = V / A_w \qquad \text{(eq 4-2)}$$

where

f_v = unit shear stress

V = vertical shear due to external loads

A_w = area of web

d. Rotational force. A rotational force or moment exerts a turning or bending effect on a member in the plane of the member and may be considered as a force acting at the end of a rigid stick. The units of a moment are the product of a force and a distance (or arm). These may be pound-inches, pound-feet, kip-inches, kip-feet, etc. When an external moment is applied to a beam or member, an internal resisting moment is developed. This internal moment is formed by longitudinal compressive and tensile fiber stresses throughout the beam, acting about the neutral axis. This action is illustrated in figure 4-16. Note that the stresses are greatest at the upper and lower beam surfaces and decline linearly to zero at the neutral axis. The maximum flexural (or bending, or fiber) stress are calculated by the following formula:

$$f_b = M c / I \qquad \text{(eq 4-3)}$$

where

f_b = bending stress

M = moment

c = distance from neutral axis to extreme fiber (or surface) of beam

I = moment of inertia

For stresses at points between the neutral axis and the extreme fiber, use the distance from the neutral axis to the point in question rather than "c." While bending occurs in many structures, it is most common in beam and girder spans. The most common use of a beam is in a simple span. A simple span could be a timber, concrete, or steel beam supported on abutments at each end. The dead and live loads cause downward forces which, with the reactions form external moments, result in the bending of the beam between its supports. The bending produces compressive stresses in the upper, or concave, portion of the beam and tensile stresses in the lower or convex portion of the beam (figure 4-17). A moment producing this type of bending is considered positive. Positive moment is typical of vertical loads acting on simple beams.

e. Negative movement. A continuous (over intermediate supports) beam is shown in figure 2-1 (part b). It is apparent that the same type of loading will produce a positive moment acting on the middle of the span. However, over the support, the upper part of the beam will elongate while the lower part will shorten. This is called negative moment and is present in continuous structures. A negative moment can also be produced in a simple beam (figure 2-1, part a> by an uplift force.

f. Vertical loads. In addition to these horizontal fiber stresses, the vertical loads on the structure are carried to the reactions at the span ends by means of shearing stress in the web of the beam. The beam must, of course, be sized so that all the stresses which it is to withstand will be within allowable limits. It is also important that the beam be rigid enough to keep its deflection within proper limits even when the stresses do not approach limiting values.

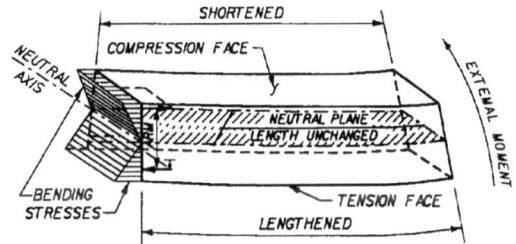

Figure 4-16. Bending stress in a beam.

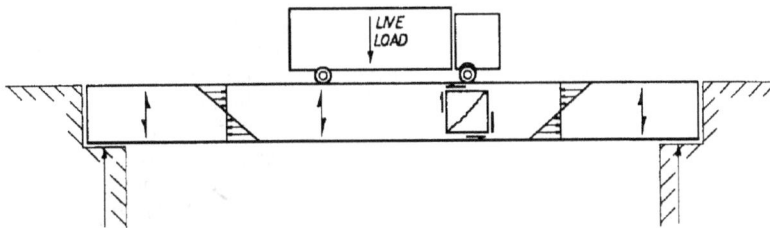

Figure 4-17. Simple beam bending moment and shear.

CHAPTER 5

BRIDGE CONSTRUCTION MATERIALS

Section I. CONCRETE

5-1. General

Concrete is essentially a compressive material. While it has adequate strength for most structural uses, it is best suited for relatively massive members that transmit compressive loads directly to the founding material. Although concrete has low tensile strength, reinforcing it with steel bars produces a material that is suitable for the construction of flexural members, such as deck slabs, bridge girders, etc. Prestressed concrete is produced by a technique which applies compression to concrete by means of highly stressed strands and bars of high strength steel wire. This compressive stress is sufficient to offset the tensile stress caused by the applied loads. Prestressing has greatly increased the maximum span length of concrete bridges.

5-2. Physical and mechanical properties

a. *Strength.* Compressive strength is high, but shear and tensile strengths are much lower, being about 12 percent and 10 percent, respectively, of the compressive strength.

b. *Porosity.* Concrete is inherently porous and permeable since the cement paste never completely fills the spaces between the aggregate particles. This permits absorption of water by capillary action and the passage of water under pressure.

c. *Extensibility.* Concrete is considered extensible, i.e., undergoes large extensions without cracking. However, this presupposes a high-quality concrete and freedom from restraint.

d. *Fire resistance.* High-quality concrete is highly resistant to the effects of fire. However, intense heat will damage concrete.

e. *Elasticity.* Concrete under ordinary loads is elastic, i.e., stress is proportional to strain. Under sustained loads, the elasticity of concrete is significantly lowered due to creep. This makes concrete less likely to crack.

f. *Durability.* The durability of concrete is affected by climate and exposure. In general, as the water-cement ratio is increased, the durability will decrease correspondingly. Properly proportioned, mixed, and placed, concrete is very durable.

g. *Anisotropy.* Concrete itself is generally isotropic, but once reinforced with steel bars or pre-

stressed with steel wires, it becomes anisotropic, i.e., its strength varies depending on the direction in which it is loaded.

5-3. Indication and classification of deterioration

While performing an inspection of concrete structures, it is important that the conditions observed be described in very clear and concise terms that can later be understood by others. The common terms used to describe concrete deterioration are discussed:

a. *Cracking.* Cracks in concrete may be described in a variety of ways. Some of the more common ways are in terms of surface appearance, depth of cracking, width of cracking, current state of activity, and structural nature of the crack:

(1) *Surface appearance.* The surface appearance of cracks can give the first indication of the cause of cracking. Two categories exist:

(a) *Pattern or map cracks.* These are rather short cracks, usually uniformly distributed and interconnected, that run in all directions (figure 5-1). These cracks are the result of restraint of contraction of the surface layer or possibly an increase of volume in the interior of the concrete. Another type of pattern crack is "D-cracking." These cracks usually start in the lower part of a concrete slab adjacent to joints, where moisture accumulates and progresses away from the corners of the slab (figure 5-2). Vertical cracks near vertical expansion joints in abutments and walls can also be classified as D-cracks.

(b) *Individual cracks.* These cracks run in definite directions and may be multiple cracks in parallel at definite intervals. Individual cracks indicate tension in the direction perpendicular to the cracking. Several terms may be used to describe the direction that an individual or isolated crack runs: diagonal, longitudinal, transverse, vertical, and horizontal. The directions of these cracks are demonstrated in figure 5-3.

(2) *Depth of cracking.* This category is self-explanatory. The four categories generally used to describe crack depth are surface, shallow, deep, and through.

(3) *Width of cracking.* Three width ranges are used: fine (generally less than 1/32 inch); medium

Figure 5-1. Pattern or map cracking on a pier.

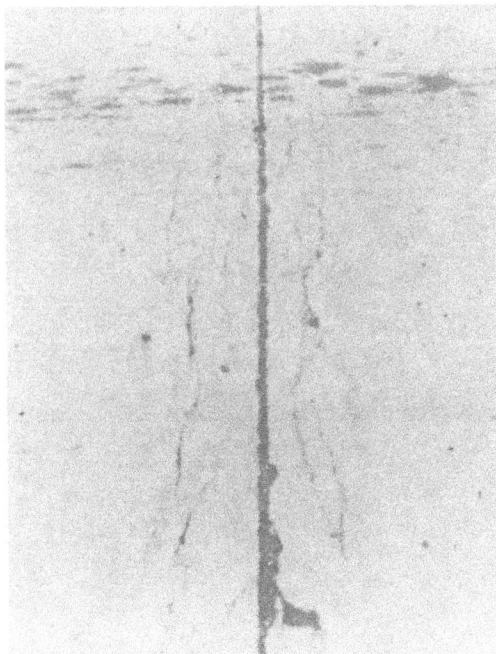

Figure 5-2. D-cracking on a deck.

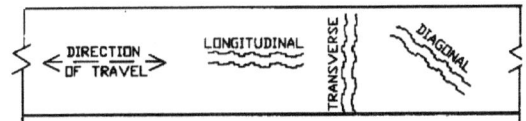

PLAN VIEW OF A BRIDGE DECK

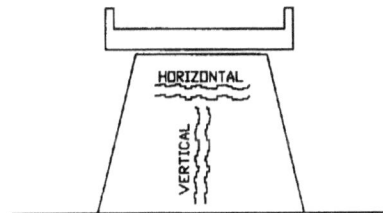

ELEVATION VIEW OF A PIER

Figure 5-3. Nomenclature for individual cracks.

(between 1/32 and 1/16 inch); and wide (over 1/16 inch).

(4) *Current state of activity.* The activity of the crack refers to the presence of the factor causing the cracking. The activity must be taken into account when selecting a repair method. Two categories exist:

(a) *Active cracks.* These are cracks for which the mechanism causing the cracking is still at work. If the crack is currently moving, regardless of why the crack formed initially or whether the forces that caused it to form are or are not still at work, it must be considered active. Also, any crack for which an exact cause cannot be determined should be considered active.

(b) *Dormant cracks.* These are cracks which are not currently moving or for which the movement is of such magnitude that a repair material will not be affected by the movement.

(5) *Structural nature of the crack.* Cracks may also be categorized as structural (caused by excessive live or dead loads) and nonstructural (caused by other means). Structural cracks will usually be substantial in width, and the opening may tend to increase as a result of continuous loading and creep of the concrete. In general, it can be difficult to determine readily during a visual examination whether a crack is structural or nonstructural. Such a determination will frequently require a structural engineer.

(6) *Combination* of *descriptions.* To describe cracking accurately, it will usually be necessary to use several terms from the various categories listed. For example: (1) shallow, fine, dormant, pattern cracking, or (2) shallow, wide, dormant, isolated short cracks.

b. Disintegration. Disintegration of concrete may be defined as the deterioration of the concrete into small fragments or particles due to any cause. It differs from spalling in that larger pieces of intact concrete are lost when spalling occurs. Disintegration may be caused by a variety of causes including aggressive water attack, freezing and thawing, chemical attack, and poor construction practices. Two of the most commonly used terms used to describe disintegration are scaling and dusting:

(1) *Scaling.* This is the gradual and continuing loss of surface mortar and aggregate over an area. The inspector should describe the character of the scaling, the approximate area involved, and the location of the scaling on the bridge. Scaling should be classified as follows:

(a) *Light scale.* Loss of surface mortar up to ¼ inch deep, with surface exposure of coarse aggregates (figure 5-4), is considered light scale.

(b) *Medium scale.* Loss of surface mortar from ¼ to ½ inch deep, with some added mortar loss between the coarse aggregates (figure 5-5), is considered medium scale.

(c) *Heavy scale.* Loss of surface mortar surrounding aggregate particles of ½ to 1 inch deep is considered heavy scale. Aggregates are clearly exposed and stand out from the concrete (figure 5-6).

(d) *Severe scale.* Loss of coarse aggregate particles as well as surface mortar and the mortar surrounding the aggregates is considered severe scale. Depth of the loss exceeds 1 inch.

Figure 5-4. Light scale.

Figure 5-5. Medium scale.

Figure 5-6. Heavy scale.

(2) *Dusting.* Dusting is the development of a powdered material at the surface of hardened concrete. Dusting will usually be noted on horizontal concrete surfaces that receive a great deal of traffic. Typically, dusting is a result of poor construction practice. For example, sprinkling water on a concrete surface during finishing will frequently result in dusting.

c. *Spalling.* Spalling is defined as the development of fragments, usually in the shape of flakes, detached from a larger mass. As previously noted, spalling differs from disintegration in that the material being lost from the mass is concrete and not individual aggregate particles that are lost as the binding matrix disintegrates. The distinction between these two symptoms is important when attempting to relate symptoms to causes of concrete problems. Spalls can be categorized as follows:

(1) *Small spall.* These are not greater than ¾ inch in depth nor greater than 6 inches in any dimension (figure 5-7).

(2) *Large spall.* These are deeper than ¾ inch and greater than 6 inches in any dimension (figure 5-8).

(3) *Special case of spalling.* Two special cases of spalling must be noted:

(a) *Popouts.* These appear as shallow, typically conical depressions in a concrete surface (figure 5-9). They may be the result of freezing of concrete that contains some unsatisfactory aggregate particles. They are easily recognizable by the shape of the pit remaining in the surface and by a

Figure 5-7. Small spall.

portion of the offending aggregate particle usually being visible in the hole.

(b) *Spalling caused by the corrosion of reinforcement.* One of the most frequent causes of spalling is the corrosion of reinforcing steel. During a visual examination, spalling caused by corrosion of reinforcement is usually an easy symptom to recognize since the corroded metal will be visible along with rust staining, and the diagnosis will be straightforward (figure 5-10).

(c) *Joint spall.* This is an elongated depression along an expansion, contraction, or construction joint (figure 5-11).

5-4. Causes of deterioration

Once the inspection of a concrete structure has been completed, the cause or causes for any deterioration must be established. Since many of the symptoms may be caused by more than one mechanism acting upon the concrete, it is necessary to have an understanding of the basic underlying causes of damage and deterioration. Table 5-1 summarizes the various causes of deterioration in concrete and their associated indicators. These causes are discussed:

a. *Accidental loadings.* These loadings are generally short-duration, one-time events such as vehicular impact or an earthquake. These loading:

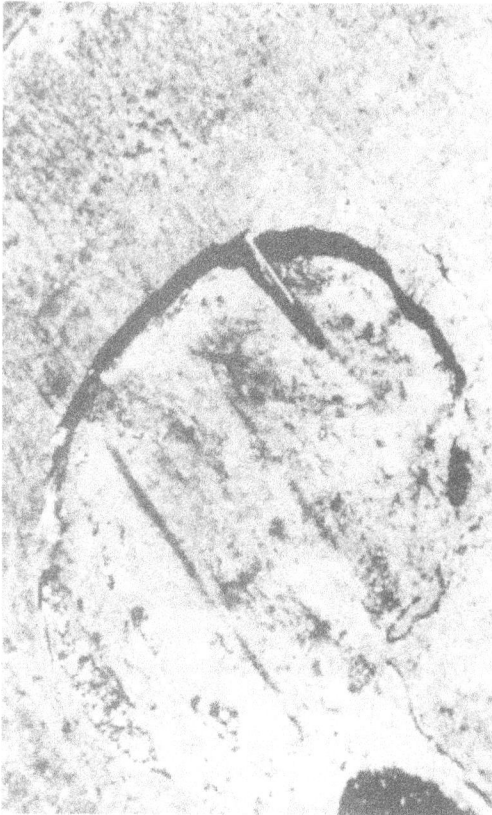

Figure 5-8. Large spall.

Figure 5-9. Popouts.

can generate stresses higher than the strength of the concrete resulting in localized or general failure. This type of damage is indicated by spalling or cracking of the concrete. Laboratory analysis is generally not necessary.

b. Chemical *reactions*. This category includes several specific causes of deterioration that exhibit a wide variety of symptoms as described:

(1) *Acid attack*. Portland cement is generally not very resistant to attack by acids, although weak acids can be tolerated. The products of combustion of many fuels contain sulfurous gases which combine with moisture to form sulfuric acid.

Figure 5-10. Spall due to reinforcement corrosion.

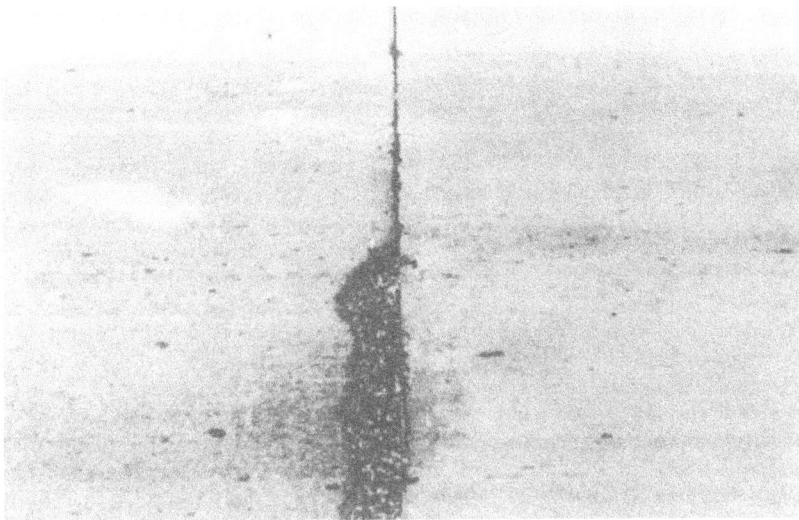

Figure 5-11. Joint spall.

Table 5-1. Relation of symptoms to causes of distress and deterioration of concrete

Symptoms	Construction Faults	Cracking	Disintegration	Distortion/ Movement	Erosion	Joint Failures	Seepage	Spalling
Accidental loadings		X						X
Chemical reactions		X	X				X	
Construction errors	X	X				X	X	X
Corrosion		X						X
Design errors		X				X	X	X
Erosion			X		X			
Freezing and thawing		X	X					X
Settlement and movement		X		X		X		
Shrinkage		X						
Temperature changes		X				X		X

Other possible sources for acid formation are sewage, some peat soils, and some mountain water streams. Visual examination will show disintegration of the concrete leading to the loss of cement paste and aggregate from the matrix. If reinforcing steel is reached by the acid, rust staining, cracking, and spalling may be seen. If the nature of the solution in which the deteriorated concrete is located is unknown, laboratory analysis can be used to identify the specific acid involved.

(2) *Alkali-carbonate rock reaction.* Certain aggregates of carbonate rock have been reactive in

concrete. The results of these reactions have been both beneficial and destructive. Visual examination of those reactions that are serious enough to disrupt the concrete in a structure will generally show map or pattern cracking and a general appearance which indicates swelling of the concrete. This reaction is distinguished from that of the alkali-silica reaction by the lack of silica gel exudations at cracks. Petrographic examination may be used to confirm the presence of alkali-carbonate rock reaction.

(3) *Alkali-silica reaction.* Some aggregates containing silica that is soluble in highly alkaline solutions may react to form expansive products which will disrupt the concrete. This reaction is indicated by map or pattern cracking and a general appearance of swelling of the concrete. Petrographic examination may be used to confirm the presence of this reaction.

(4) *Miscellaneous chemical attack.* Concrete will resist chemical attack to varying degrees depending upon the exact nature of the chemical. Concrete which has been subjected to chemical attack will usually show surface disintegration and spalling and the opening of joints and cracks. There may also be swelling and general disruption of the concrete mass. Aggregate particles may be seen protruding from the remaining concrete mass.

(5) *Sulfate attack.* Naturally occurring sulfates of sodium, potassium, calcium, or magnesium are sometimes found in soil or in solution in ground water adjacent to concrete structures. The reactions involving these sulfates result in an increase in volume of the concrete. Visual inspection will show map and pattern cracking as well as a general disintegration of the concrete. Laboratory analysis can verify the occurrence of the reactions described.

c. *Construction errors.* Failure to follow specified procedures and good practice or outright carelessness may lead to a number of conditions that may be grouped together as construction errors. Typically, most of these errors do not lead directly to failure or deterioration of concrete. Instead, they enhance the adverse impacts of other mechanisms identified in this chapter. The following are some of the most common construction errors:

(1) *Addition of water to concrete.* The addition of water while in the delivery truck will often lead to concrete with reduced strength and durability. The addition of water while finishing a slab will cause crazing and dusting of the concrete in service.

(2) *Improper consolidation.* Improper consoli-

dation of concrete may result in a variety of defects, the most common being bugholes, honeycombing, and cold joints. These defects make it much easier for any damage-causing mechanism to initiate deterioration of the concrete.

(3) *Improper curing.* Unless concrete is given adequate time to cure at a proper humidity and temperature, it will not develop the characteristics that are expected and that are necessary to provide durability. Symptoms of improperly cured concrete can include various types of cracking and surface disintegration. In extreme cases where poor curing leads to failure to achieve anticipated concrete strengths, structural cracking may occur.

(4) *Improper location of reinforcing steel.* This section refers to reinforcing steel that is either improperly located or is not adequately secured in the proper location. Either of these faults may lead to two general types of problems. First, the steel may not function structurally as intended resulting in structural cracking or failure. The second type of problem stemming from improperly located or tied steel is one of durability. The tendency seems to be for the steel to end up close to the surface of the concrete. As the concrete cover over the steel is reduced, it is much easier for corrosion to begin.

d. *Corrosion of embedded metals.* Under most circumstances, Portland-cement concrete provides good protection to the embedded reinforcing steel. This protection is generally attributed to the high alkalinity of the concrete adjacent to the steel and to the relatively high electrical resistivity of the concrete. However, this corrosion resistance will generally be reduced over a long period of time by carbonation, and the steel will begin to corrode. Deicing salts are the most common cause of the corrosion. Corrosion of the steel will cause two things to occur. First, the cross-sectional capacity of the reinforcement is reduced which in turn reduces the load-carrying capacity of the steel. Second, the products of the corrosion expand since they occupy about eight times the volume of the original material. This leads to cracking and ultimately spalling of the concrete. For mild steel reinforcing, the damage to the concrete will become evident long before the capacity of the steel is reduced enough to affect its load-carrying capacity. However, for prestressing steel slight reductions in section can lead to catastrophic failure. Visual examination will typically reveal rust staining of the concrete. This staining will be followed by cracking. Cracks produced by corrosion usually run in straight, parallel lines at uniform intervals corresponding to the spacing of the reinforcement. As deterioration continues, spalling of the concrete over the reinforcing steel will occur with the reinforcing bars becoming visible. A laboratory analysis may be beneficial to determine the chloride contents in the concrete throughout its depth. This procedure may be used to determine the amount of concrete to be removed during a rehabilitation project.

e. Design errors. Design errors generally result from inadequate structural design or from lack of attention to relatively minor design details:

(1) *Inadequate structural design.* This will cause cracking and/or spalling in areas which are subject to the highest stresses. To identify this as a source of damage, the locations of the damage should be compared to the types of stresses that should be present in the concrete. A detailed structural analysis may be required, and thus a qualified structural engineer should be consulted if this problem is apparent.

(2) *Poor design details.* Poor detailing may result in localized concentrations of high stresses in otherwise satisfactory concrete. The following are some of the more design detail problems:

(a) *Abrupt changes in section.* This may cause stress concentrations that may result in cracking. Typical examples would include the use of relatively thin bridge decks rigidly tied into massive abutments or patches and replacement concrete that are not uniform in plan dimensions.

(b) *Reentrant corners and openings.* These locations are subject to stress concentrations, and when insufficiently reinforced, cracking may occur.

(c) *Inadequate drainage.* This will cause ponding of water, which may result in excessive loading or, more likely, leakage or saturation of concrete. Concrete subject to freeze-thaw cycles is especially vulnerable to this type of damage.

(d) *Insufficient travel in expansion joints.* Inadequately designed expansion joints may result in spalling of concrete adjacent to the joints.

(e) *Rigid joints between precast units.* Designs utilizing precast elements must provide for movement between adjacent precast elements or between the precast elements and the supporting frame. Failure to provide for this movement can result in cracking or spalling.

f. Wear and abrasion. Traffic abrasion and impact cause wearing of bridge decks; while curbs, parapets, and piers are damaged by the scraping action of such vehicles as snow plows and sweepers. Deck wear also appears as cracking and ravelling at joint edges.

g. Freezing and thawing. The cyclic freezing and thawing of critically saturated concrete will cause its deterioration. Deicing chemicals may also accelerate the damage and lead to pitting and scaling. This damage ranges from surface scaling to extensive disintegration. Laboratory examination of concrete cores with this damage will often show a series of cracks parallel to the surface of the structure.

h. Foundation movement. These movements will cause serious cracking in structures. Further discussion of this problem is provided in section VII of this chapter.

i. Shrinkage. Shrinkage is caused by the loss of moisture from concrete. It may be divided into two categories: that which occurs before setting (plastic shrinkage) and that which occurs after setting (drying shrinkage). Cracking due to plastic shrinkage will be seen within a few hours of concrete placement. The cracks are generally wide and shallow and isolated rather than patterned. Cracks due to drying shrinkage are characterized by their fineness and absence of any indication of movement. They are usually shallow, a few inches in depth, and in an orthogonal or blocky pattern.

j. Temperature changes. Changes in temperature cause a corresponding change in volume of the concrete, and when sufficiently restrained against expansion or contraction cracking will occur. Temperature changes will generally result from the heat of hydration of cement in large concrete placements, variations in climatic conditions, or fire damage.

5-5. Assessment of concrete

Assessment of existing reinforced concrete in bridges is basically associated with fully identifying the cause and extent of observed or suspected deterioration. The method of description of damaged concrete and the possible causes for deterioration were discussed in paragraph 5-5. This section will provide guidance concerning the available test methods for determining the causes of the deterioration and quantifying its extent. The range of available test methods is large and includes in situ nondestructive tests upon the actual structure as well as physical, chemical, and petrographic tests upon samples removed from the structure, and load testing. Table 5-2 summarizes basic characteristics of the most widely established test methods classified according to the features which may be assessed most reliably in each case. An overview of some of the more common tests follows.

Table 5-2. Test methods for concrete

Property Under Investigation	Test	ASTM Designation	Equipment Type
Corrosion of embedded steel	Half-cell potential	C876	Electrical
	Resistivity		Electrical
	Cover depth		Electromagnetic
	Carbonation depth		Chemical and microscopic
	Chloride penetration		Chemical and microscopic
Concrete quality, durability and deterioration	Rebound hammer	C805	Mechanical
	Ultrasonic pulse velocity	C597	Electronic
	Radiography		Radioactive
	Radiometry		Radioactive
	Permeability		Hydraulic
	Absorption		Hydraulic
	Petrographic	C856	Microscopic
	Sulphate content		Chemical
	Expansion		Mechanical
	Air content	C457	Microscopic
	Cement type and content	C85, C1084	Chemical and microscopic
Concrete strength	Cores	C42, C823	Mechanical
	Pullout	C900	Mechanical
	Pulloff		Mechanical
	Breakoff		Mechanical
	Internal fracture		Mechanical
	Penetration resistance	C803	Mechanical
Integrity and structural performance	Tapping		Mechanical
	Pulse-echo		Mechanical/electronic
	Dynamic response	C215	Mechanical/electronic
	Thermography		Infrared
	Strain or crack measurement		Optical/mech./elec.

Additional references should be consulted prior to actual usage of these tests.

a. Core drilling. Core drilling to recover concrete for laboratory analysis or testing is the best method of obtaining information on the condition of concrete within a structure. However, since core drilling is expensive and destructive, it should be considered only when sampling and testing of interior concrete is deemed necessary. The core samples should be sufficient in number and size to permit appropriate laboratory examination and testing. For compressive strength, static or dynamic modulus of elasticity, the diameter of the core should not be less than three times the nominal maximum size of aggregate. Warning should be given against taking NX size (2 1/8-inch diameter) cores in concrete containing 2- to 6-inch maximum size aggregate. Due to the large aggregate size, these cores will generally be recovered in short broken pieces. When drilling in poor-quality concrete with any size core barrel, the material will generally come out as rubble. When drill hole coring is not practical or core recovery is poor, a viewing system such as a borehole camera, borehole television, or borehole televiewer may be used for evaluating the interior concrete conditions. In addition, some chemical tests may be performed on smaller drilled powdered samples from the structure, thus causing substantially less damage than that produced by coring, but the likelihood of sample contamination is increased and precision may be reduced.

b. Laboratory investigations. Once samples of concrete have been obtained, whether by coring, drilling, or other means, they should be examined in a qualified laboratory. In general, the examination will include one or more of the following examinations:

(1) *Petrographic examination.* This type of examination may include visual and microscopic inspection, x-ray diffraction analysis, differential thermal analysis, x-ray emission techniques, and thin section analysis. These techniques may be expected to provide information on the following: aggregate condition; pronounced cement-aggregate reactions; deterioration of aggregate particles in place; denseness of cement paste; homogeneity of the concrete; depth and extent of carbonation; occurrence and distribution of fractures; characteristics and distribution of voids; and presence of contaminating substances.

(2) *Chemical analysis.* Chemical analyses of hardened concrete or of selected portions (paste, mortar, aggregate, reaction products, etc.) may be used to estimate the cement content, original water-cement ratio, and the presence and amount of chloride and other admixtures. The chloride analysis is the most common of these analyses. It is used to provide a quantitative measure of chloride ion contamination and, thus, the potential for active steel corrosion at various levels in the concrete deck. Samples for this test are usually taken by a rotary hammer drill. The "threshold" chloride content, or amount of chloride needed to initiate corrosion, is approximately 2.0 pounds of chloride content per cubic yard of concrete.

(3) *Physical analysis.* The following physical and mechanical tests are generally performed on concrete cores: density, compressive strength, modulus of elasticity, Poisson's ratio, pulse velocity, and volume change potential by freezing and thawing.

c. Nondestructive testing (NDT). The purpose of NDT is to determine the various properties of concrete such as strength, modulus of elasticity, homogeneity, and integrity, as well as conditions of strain and stress, without damaging the structure. Some of the most commonly used tests are discussed:

(1) *Rebound number (hammer).* Rebound numbers may be used to estimate the uniformity and quality of in situ concrete. The rebound number is obtained by the use of a special "hammer" that consists of a steel mass and a tension spring in a tubular frame. The measured rebound number can be related to calibration curves which will give an indication of the in situ concrete strength. The rebound number increases with the strength of the concrete. This method is inexpensive and allows for a large number of measurements to be rapidly taken so that large exposed areas of concrete can be mapped within a few hours. It is, however, a rather imprecise test and does not provide a reliable prediction of the strength of concrete. The measurements can be affected by: smoothness of the concrete surface; moisture content of the concrete; type of coarse aggregate; size, shape, and rigidity of the specimen; and carbonation of the concrete surface.

(2) *Penetration resistance (probe).* This test is also used for a quick assessment of quality and uniformity of concrete. The apparatus most often used for penetration resistance is the Windsor Probe, a special gun which uses a 32-caliber blank with a precise quantity of powder to fire a high-strength steel probe into the concrete. The depth of penetration of the probe into the concrete can then be related by calibration curves to concrete compressive strength. A probe will penetrate deeper as the density, subsurface hardness, and strength of the concrete decrease. It should not be considered for use as a precise measurement of concrete strength. However, useful estimates of the compressive strength may be obtained if the probe is properly calibrated. This test does damage the concrete, leaving a hole of about 0.32 inches in diameter for the depth of the probe, and may cause minor cracking and some surface spalling. Minor repairs of exposed surfaces may be necessary.

(3) *Ultrasonic pulse velocity.* This method involves the measurement of the time of travel of electronically pulsed compressional waves through a known distance in concrete. These velocities can be used to assess the general condition and quality of concrete, to assess the extent and severity of cracks in concrete,   and to delineate areas of deteriorated or poor-quality concrete. Good-quality, continuous concrete will normally produce high velocities accompanied by good signal strengths. Poor-quality or deteriorated concrete will usually decrease velocity and signal strength. Concrete of otherwise good quality, but containing cracks, may produce high or low velocities, depending upon the nature and number of cracks, but will almost always diminish signal strength. This method does not provide a precise estimate of concrete strength. Moisture variations and the presence of reinforcing steel can affect the results. Skilled personnel are required in the analysis of the results.

(4) *Surface tapping (chain drag).* Experience has shown that the human ear, used in conjunction with surface tapping, is the most efficient and economical method of determining major delamination in bridge decks. Chain dragging is the most commonly used method for this purpose. This is, however, a very subjective test in that the operator must be able to differentiate between sound and unsound regions, and the results cannot be easily quantified.

d. Steel corrosion assessment. The most commonly used test for assessing the current state of reinforcing steel corrosion is the half-cell potential test. This test involves measurement of the electrical potential of an embedded reinforcing bar relative to a reference half-cell placed on the concrete surface. Potential differences more negative than - 0.35 volts indicates a high degree of probability of active corrosion of the reinforcing steel. Potential readings of - 0.20 volts and lower indicate the probability of inactive or no corrosion, while readings between -0.20 and -0.35 volts indicate the possibility of active corrosion.

e. Load tests. Occasionally it may be necessary to examine the overall behavior of an entire bridge structure or section of a bridge. This may be achieved electronically by measuring the response to dynamic loading with the aid of appropriately positioned accelerometers or alternatively monitoring the performance under static test loads. The most common method however, is to measure strains and deflections (with precise leveling or lasers) produced from static full-scale test loads. These tests are generally expensive but yield valuable information as to the overall "health" of a structure. This type of test can be conducted on any type of bridge, regardless of the material type.

Section II. STRUCTURAL STEEL

5-6. Physical and mechanical properties

a. Strength. Steel possesses tremendous compressive and tensile strength and is highly resistant to shear forces. Thin steel sections, however, are vulnerable to compressive buckling.

b. Ductility. Both the low-carbon and low-alloy steels normally used in bridge construction are quite ductile. Brittleness may occur because of heat treatment, welding, or through metal fatigue.

c. Durability. Steel, when protected properly, is extremely durable.

d. Fire resistance. Steel is subject to a loss of strength when exposed to high temperatures such as those resulting from fire.

e. Corrosion. Unprotected carbon steel corrodes (rusts) readily. However, it can readily be protected.

f. Weldability. Although steel is weldable, it is necessary to determine the chemistry of the steel and to select a suitable welding procedure before starting welding operations on a bridge.

g. Others. Steel is elastic and conducts heat and electricity.

5-7. Indicators and classification of deterioration

a. Rust. Rusted steel varies in color from dark red to dark brown. Initially, rust is fine grained, but as it progresses it becomes flaky or scaly in character. Eventually, rust causes a pitting of the member. The inspector should note the location, characteristics, and the extent of the rusted areas. The depth of heavy pitting should be measured and the size of any perforation caused by rusting should be recorded. Rust may be classified as follows:

(1) *Light.* A light, loose rust formation pitting the paint surface.

(2) *Moderate.* A looser rust formation scales or flakes forming. Definite areas of rust are discernible.

(3) *Severe.* A heavy, stratified rust or rust scale with pitting of the metal surface. This rust condition eventually culminates in the perforation of the steel section itself.

b. Cracks.

(1) Cracks in the steel may vary from hairline thickness to sufficient width to transmit light through the member. The first visible evidence is normally a crack in the paint film. Depending on the location and the length of time the paint has been open, there may be a thin line of rust stain emanating from the crack as shown in figure 5-12. Crack identification on unpainted A588 steel is particularly difficult. There is no staining due to oxidation and the rough surface texture tends to hide the crack.

(2) Any type of crack is obviously serious and should be reported at once. Record the location and length of all cracks and indicate whether the cracks are open or closed. The full length of the crack may not be completely visible. A suitable nondestructive test such as the dye penetrant test (discussed in the following section) can help to establish its full length.

(3) Cracks in fracture critical steel members (which should have been identified prior to the inspection) are especially serious and therefore these members should be inspected with extreme care. Cracks of any size should be immediately reported to the appropriate authority.

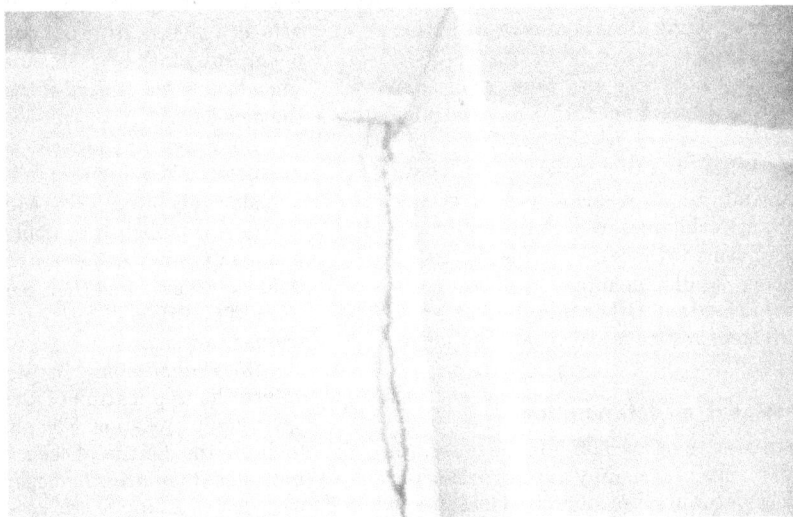

Figure 5-12. Rust stained crack in steel girder.

c. Buckles and kinks. These conditions develop mostly because of damage arising from thermal strain, overload, or added load conditions. The latter condition is caused by failure or the yielding of adjacent members or components. Collision damage may also cause buckles, kinks, and cuts (figure 5-13). Look for cracks radiating from cuts or notches. Note the members damaged, the type, location, and extent of the damage, and measure the amount of deformation, if possible.

d. Stress concentrations. Observe the paint around the connections at joints for fine cracks as indications of large strains due to stress concentrations. Be alert for sheared or deformed bolts and rivets.

e. Galvanic corrosion. This condition will appear essentially similar to rust.

Figure 5-13. Buckled flange due to collision.

5-8. Causes of deterioration

a. Air and moisture. Air and moisture cause rusting of steel, especially in a marine climate.

b. Industrial fumes. Industrial fumes in the atmosphere, particularly hydrogen sulfide, cause deterioration of steel.

c. Deicing agents. While all deicing agents attack steel, salt is the most commonly encountered chemical on bridges.

d. Seawater and mud. Unprotected steel members, such as piles immersed in water and embedded in mud, undergo serious deterioration and loss of section.

e. Thermal strains or overloads. Where movement is restrained, or where members are overstressed, the steel may yield, buckle, or crack (or rivets and bolts may shear).

f. Fatigue and stress concentrations. Cracks may develop because of fatigue or poor details which produce high stress concentrations. Examples of such details are: reentrant corners, abrupt and large changes in plate widths and/or thicknesses, a concentration of heavy welds, or an insufficient bearing area for a support. Fatigue and stress concentrations are very important factors in the failure of steel structures. A thorough discussion of these factors is provided in the American Association of State Highway Transportation Officials (AASHTO) Manual, "Inspection of Fracture Critical Bridge Members."

g. Fire. Extreme heat will cause serious deformations of steel members.

h. Collisions. Trucks, over-height loads, derailed cars, etc., may strike steel beams or columns, damaging the bridge.

i. Animal wastes. These may cause rusting and can be considered as a special type of direct chemical attack.

j. Welds. Where the flux is not neutralized, some rusting may occur. Welds may crack because of poor welding techniques or poor weldability of the steel. Problems with welds will also be discussed in more detail in chapter 6.

k. Galvanic action. Other metals that are in contact with steel may cause corrosion similar to rust.

5-9. Assessment of deterioration

a. As discussed in paragraph 5-9, the deterioration of steel members is mainly due to rusting or cracking. The determination of section loss due to rusting is of primary importance for a proper load rating assessment to be performed. This is usually done with precise mechanical (such as calipers) or electrical (such as a "depth meter") measuring devices. The full assessment of cracks in steel members is very important since they can cause rapid failure to some members. All of the cracks in a member should be located and their extent of propagation should be fully defined. In addition to a close visual examination, a wide variety of nondestructive test methods exists for these purposes and several of the most common methods are briefly discussed:

(1) *Dye penetrant.* This test is used to identify the location and extent of surface cracks and surface defects, such as hairline fatigue cracks. It cannot be used to locate subsurface defects. For the test, the area must be thoroughly cleaned of paint, rust, scale, grease, and oily films. Then, an oil-based liquid penetrant is applied which is intended to be absorbed into any cracks present. After a specific amount of time, the excess is wiped off and a developer applied. The developer acts as a blotter, drawing out a portion of the penetrant which has seeped into the defect, causing a bright red outline of the defect to appear in the developer.

(2) *Magnetic particle testing.* In this test, a magnetic field is induced in the steel by means of a moderately sized power source. Detection of a flaw is accomplished by application of inert compounds of iron which are attracted to the magnetic field as it leaves and then reenters the steel in the area of the flaw. This test requires a highly trained inspector.

(3) *Radiographic inspection.* This process involves the application of x rays to an area or specimen in question. The ability of the specimen to dilute the density of the x rays passing through indicates its relative homogeneity. Any discontinuity, such as a fatigue crack, will show up on film placed behind the specimen as less dense than the sound material. This test method is most beneficial and has been used most successfully in analyzing welds for incomplete fusion, slag and other inclusions, incomplete penetration, and gas pockets.

b. Tensile coupons. To perform an accurate analysis of the bridge's load capacity, the material properties of the steel must be known. In many older bridges, the type of steel, and thus its properties, may not be known. In these cases, the cutting of "coupons" for testing may be necessary. Since this operation causes considerable damage to the member from which the coupon is taken, extreme care must be exercised. The location for the coupon should be carefully chosen to provide the most accurate information while causing the least structural damage to the structure. Guidance from a structural engineer should be obtained. The samples should be 9 to 12 inches long and 2 to 3 inches wide. The actual coupons will be machined from these samples.

Section III. TIMBER

5-10. Physical and mechanical properties

a. Strength. Timber, while not as strong as steel, approximates ordinary concrete in compressive strength. Rated strongest in flexural strength, timber has an allowable compressive strength (parallel to grain) of about 75 percent of the flexural value. Perpendicular to the grain, compressive strength is only 20 percent of the flexural strength. Horizontal shear is limited to 10 percent of the flexural strength.

b. Porosity. Being a cellular, organic material, timber is quite porous.

c. Anisotropy. As may be deduced from the differences in allowable compressive strengths, wood is anisotropic, i.e., it has different strength properties depending upon the manner and direction of loading.

d. Impact resistance. Since timber is able to withstand a greatly increased load momentarily, neither impact nor fatigue are serious problems with timber.

e. Durability. Under certain conditions, and when properly treated or protected, timber is quite durable. However, timber is not a particularly durable material under all conditions. It should be noted that some preservative treatments reduce the strength of timber.

f. Fire resistance. Timber is very vulnerable to damage by fire.

g. Other. Timber is also elastic, low in thermal and electrical conductivity, and subject to volume changes.

5-11. Deterioration: indicators and causes

a. Fungus decay. Fungi usually require some moisture to exist. As a rule, fungus decay can be avoided only by excellent preservative treatment. Fungus decay is classified as follows:

(1) *Mild.* Mild fungus decay appears as a stain or discoloration. It is hard to detect and even harder to distinguish between decay fungi and staining fungi.

(2) *Advanced.* Wood darkens further and shows signs of definite disintegration, with the surface becoming punky, soft and spongy, stringy, or crumbly, depending upon the type of decay or fungus (figures 5-14 and 5-15). It is similar to dryrot of door posts and outside porches. Fruiting bodies of fungi, similar to those seen on old stumps, may develop. The inspector should note the location, depth of penetration, and size of the areas of decay. Where decay occurs at a joint or splice, the effect on the strength of the connection should be indicated. A knife, icepick, or an increment borer can be used to test for decayed wood.

Decay is very likely to occur at connections, splices, support points, or around bolt holes. This may be due to the tendency of such areas to collect and retain moisture or to bolt holes or cuts being made in the surface after the preservative treatment has been applied. Unless these surfaces are subsequently protected, decay is very likely. Any holes, cuts, scrapes, or other breaks in the timber surface which would break the protective layers of the preservative treatment and allow access to untreated wood should be noted.

b. Vermin. The following vermin tunnel in and hollow out the insides of timber members for food and/or shelter:

(1) *Termites.* All damage is inside the surfaces of the wood; hence, it is not visible. White mud shelter tubes or runways extending up from the earth to the wood and on the side of masonry substructures are the only visible signs of infestation. If the timber members exhibit signs of excessive sagging or crushing, check for termite damage with an ice pick or an increment borer.

(2) *Powder-post beetles.* The outer surface is poked with small holes. Often a powdery dust is dislodged from the holes. The inside may be completely excavated.

(3) *Carpenter ants.* Accumulation of sawdust on the ground at the base of the timber is an indicator. The large, black ants may be seen in the vicinity of the infested wood.

(4) *Marine borers.* The inroads of marine borers will usually be most severe in the area between high and low water since they are waterborne, although damage may extend to the mud

Figure 5-14. Advanced wood decay in crosstie.

Figure 5-15. Advanced wood decay in bent cap.

line. Where piles are protected by concrete or metal shielding, the shields should be inspected carefully for cracks or holes that would permit entrance of the borers. Unplugged holes such as those left by test borings, nails, bolts, and the like, also permit entrance of these pests. In such cases, there are often no outside evidences of borer attack. The inspector should list the location and extent of damage and indicate whether it is feasible to exterminate the infestation and strengthen the member or if immediate replacement is necessary.

(5) *Mollusk borers (shipworms)*. The shipworm is one of the most serious enemies of marine timber installations. The most common species of shipworms is the teredo. This shipworm enters the timber in the early stage of life and remains there for the rest of its life. Teredos reach a length of 15 inches and a diameter of 3/8 inch, although some species of shipworm grow to a length of 6 feet. The teredo maintains a small opening in the surface of the wood to obtain nourishment from the seawater.

(6) *Crustacean borers*. The most commonly encountered crustacean borer is the linoria or wood louse. It bores into the surface of the wood to a shallow depth. Wave action or floating debris breaks down the thin shell of timber outside the borers' burrows, causing the linoria to burrow deeper. The continuous burrowing results in a progressive deterioration of the timber pile cross

section which will be most noticeable by the hourglass shape developed between the tide levels.

c. Weathering and warping. This is caused by repeated dimensional changes in the wood, usually due to repeated wetting. It may be described as follows:

(1) *Slight.* Surfaces of wood are rough and corrugated, and the members may even warp (figure 5-16).

(2) *Advanced.* Large cracks extend deeply or completely through the wood (figure 5-16). Wood is crumbly and obviously deteriorated.

d. Chemical attack. Chemicals act in three different ways: a swelling and resultant weakening of the wood, a hydrolysis of the cellulose by acids, or a delignification by alkalis. Animal wastes are also a problem. Chemical attack will resemble decay and should be classified similarly.

e. Fire. Timber is particularly vulnerable to fire. This type of damage is easily recognized and usually will have been reported prior to the inspection.

f. Abrasion and mechanical wear. Wear due to abrasion is readily recognized by the gradual loss of section at the points of wear (figure 5-16). This type of wear is most serious when combined with decay which softens or weakens the wood. The decks of bridges are especially vulnerable. The inspector should report the location, the general area subject to wear, and the loss in thickness. He should also indicate whether immediate remedial action is necessary.

Figure 5-16. Slight and advanced wear on a timber deck (mechanical abrasion wear also present).

g. Collision or overloading damage. Damage will be evident in the form of shattered or injured timbers, sagging or buckled members (figure 5-17), or timbers with large longitudinal cracks (figure 5-18). The inspector should give the location and extent of damage and determine whether immediate remedial action is required.

h. Unplugged holes. Holes left by test borings, nails, bolts, etc. will inevitably allow attack from many of the previously mentioned sources. Their location and extent should be noted and recommendations should be made for their repair.

5-12. Assessment of deterioration

A chipping hammer, an ice pick, and an increment borer are the primary tools used for assessment of wood deterioration. The soundness of all timbers should be first checked by tapping with the hammer and listening for a "hollow" sound. When suspect areas are found, the ice pick should be used to verify the existence of soft spots. If extensive damage is suspected, an increment borer should be used to take a test boring and fully define the extent of deterioration. Borings should be taken very selectively to not further weaken the already damaged timber, and borings should not be made at all if no deterioration is evidenced. Once the boring has been made, save the boring sample and make sure that creosote plugs are inserted into the hole made by the borer.

Figure 5-17. Buckled timber pile due to overload.

Figure 5-18. Longitudinal cracks in timber beam due to overload.

Section IV. WROUGHT AND CAST IRON

5-13. General

a. Wrought iron is a metal in which slag inclusions are rolled between the microscopic grains of iron. This results in a fibrous material with properties quite similar to steel, although tensile strength is lower than it is in steel.

b. Cast iron is iron in which carbon has been dissolved. Other elements which affect the properties of cast iron are silicon and manganese. In general, a wide range of properties are obtained depending upon the alloying elements used.

5-14. Physical and mechanical properties

a. *Wrought iron.*

(1) *Strength.* Wrought iron has an ultimate tensile strength of about 50,000 pounds per square inch. However, the rolling process and presence of the slag inclusions make wrought iron anisotropic; wrought iron has a tensile strength across the grain of about 75 percent of its longitudinal strength. While this characteristic has been largely eliminated in the wrought iron made today, wrought iron in the old bridge will be anisotropic.

(2) *Elasticity.* Wrought iron has an elastic modulus of 24,000,000 to 29,000,000 pounds per square inch. This is nearly as high as steel.

(3) *Ductility.* Wrought iron is generally ductile, although its ductility depends, to large extent, upon the method of manufacture.

(4) *Toughness and impact resistance.* Wrought iron is tough and resistant to impact.

(5) *Corrosion resistance.* The fibrous nature of wrought iron produces a tight rust which is less likely to progress to flaking and scaling than is rust on carbon steel.

(6) *Weldability.* Wrought iron is welded with no great difficulty. However, care should be exercised when planning to weld the metal of an existing bridge.

b. *Cast iron.*

(1) *Strength.* The tensile strength of cast iron varies from 20,000 to 60,000 pounds per square inch, depending upon its composition. In dealing with the iron in old bridges, it must be assumed that the cast iron will be near the lower end of the scale. The compressive strength of cast iron is high, from 60,000 pounds per square inch higher. For this reason, cast iron was used for compression members in the early iron bridges, while wrought iron was used for tension members.

(2) *Elasticity.* Cast iron has an elastic modulus of 13,000,000 pounds per square inch.

(3) *Brittleness.* Cast iron is brittle.

(4) *Impact resistance.* Cast iron possesses good impact resistant properties.

(5) *Corrosion resistance.* Cast iron, in general, is more corrosion resistant than the other ferrous metals.

(6) *Weldability.* Due to its high carbon content, cast iron is not easily welded.

5-15. Deterioration: indicators and causes

Both wrought iron and cast iron are subject to the same causes of deterioration as structural steel (previously discussed). It should be noted that cast iron is subject to defects such as checks (cracking due to tensile cooling stresses) and blowholes. The latter has a serious effect on both the strength and toughness of the cast iron.

Section V. STONE MASONRY

5-16. General

Stone masonry is little used today, except as facing or ornamentation. However, some old stone structures are still in use. The types of stones commonly used in bridges are granite, limestone, and sandstone, although many smaller bridges or culverts were built of stones locally available.

5-17. Physical and mechanical properties

a. Strength. Stone has more than adequate strength for most loads.

b. Porosity. While all stone is porous, sandstone and some limestones are much more porous than granite.

c. Absorption. Most stone is absorptive, especially limestone.

d. Thermal expansions. Stone expands and contracts with temperature variations.

e. Thermal conductivity. Stone is generally a poor conductor of heat.

f. Durability. Stone is more durable than most materials, although there is a wide range in durability between different varieties of stone.

g. Fire resistance. While not flammable, stone can be damaged by fire.

5-18. Indicators of deterioration

The following terms should be used to describe deterioration of stone masonry structures. The description, extent, and location of the deterioration should be reported.

a. Weathering. The hard surface degenerates into small granules, giving stones a smooth, rounded look.

b. Spalling. Small pieces of rock break out or chip away.

c. Splitting. Seams or cracks open up in the rocks, eventually breaking them into smaller pieces.

5-19. Causes of deterioration

a. Chemical. Gasses and solids dissolved in water often attack rocks chemically. Some of these solutions can dissolve cementing compounds between the rocks. Oxidation and hydration of some compounds found in rock will also damage.

b. Seasonal expansion and contraction. Repeated volume changes produced by seasonal expansion and contraction will cause tiny seams to develop, thereby weakening the rock.

c. Frost and freezing. Water freezing in the seams and pores of rocks can split or spall rock.

d. Abrasion. Abrasions are due mostly to wind or waterborne particles.

e. Plant growth. Lichens and ivy will attack stone surfaces chemically in attaching themselves to the stone. Roots and stems growing in crevices or joints exert a wedging force.

f. Marine borers. Rock-boring mollusks attack rock by means of chemical secretions.

Section VI. ALUMINUM

5-20. General

a. While aluminum has been widely used for signs, light standards, railings, and sign bridges, it is seldom used as the principal material in the construction of vehicular bridges. While most properties of aluminum are similar to those of steel, the following differences exist:

(1) *Lightness.* Aluminum weighs about one-third (1/3) as much as steel.

(2) *Strength.* Aluminum, while not as strong as steel, will be made comparable to steel in strength when alloyed.

(3) *Corrosion resistance.* Aluminum is highly resistant to atmospheric corrosion.

(4) *Workability.* Aluminum is easily fabricated. However, welding requires special procedures.

(5) *Durability.* Aluminum is durable.

5-21. Deterioration: indicators and causes

a. Cracking. Aluminum may be subject to some fatigue cracking. Aluminum members should be examined in areas near the bases of cantilever arms and in areas near complex welded and riveted connection. Weld cracking often occurs on bridge signs due to stresses caused from the misalignment of prefabricated sections and due to

fatigue caused from wind and vibration loadings.

b. Pitting. Aluminum will pit slightly, but this condition rarely becomes serious.

Section VII. FOUNDATION SOILS

5-22. General

a. Most foundation movements are caused by movement of the supporting soil. For this reason, it is desirable to give a brief description of these movements, although the basic theory involved is beyond the scope of this manual.

b. Soil deformations are caused by volume changes in the soil or by a shear failure. Slope slides and bearing failures are good examples of shear failures. Where loads are not large enough to cause shear failure, settlements may still occur as a result of volume change. The length of time and magnitude of the settlements depend upon the composition of the soil. Granular solids, such as sand, will usually undergo a relatively small volume change in a short period of time. However, cohesive soils such as clay can undergo large deformations or volume changes, which may continue for years. This latter process is called consolidation and is usually confined to clays and clayey silts.

c. Substructures that are supported directly by a cohesive soil may continue to settle for a long period of time. Consolidation usually produces vertical settlement.

5-23. Types of movement

For convenience, foundation movements may also be classified into the following categories:

a. Lateral movements. Earth-retaining structures, such as abutments and retaining walls, are susceptible to lateral movements, although piers sometimes also undergo such displacements.

b. Vertical movements (settlements). Any type of substructure not founded on solid rock may be subject to settlement.

c. Pile settlements. While pile settlement could be listed under lateral or vertical movements, it is mentioned separately since there is a tendency to consider piles as a panacea for all foundation problems. In addition, some of the causes of failure are peculiar to pile foundations.

d. Rotational movement (tipping). Rotation movement of substructures can be considered to be the result of unsymmetrical settlements or lateral movements. It will be discussed under the movement that is typical of the various substructures.

5-24. Effects on structures

The effects of foundation movements upon a structure will vary according to the following factors:

a. Magnitude of movements. All foundations undergo some settlement, even if only elastic compression of ledge or piles. All sizable footings probably will experience a minute differential settlement. However, very small foundation movements have no effect. Simple structures, and those with enough joints, will tolerate even moderate differential displacements with little difficulty other than minor cracking and the binding of end dams. Movements of large magnitudes, especially when differential, cause distress in nearly all structures (paragraph 5-25b(2)). Large movements will cause deck joints to jam; slabs to crack; bearings to shift; substructures to crack, rotate, or slide; and superstructures to crack, buckle, and possibly, even to collapse. The larger the settlement to be accommodated within a given distance, the more structural damage can be anticipated.

b. Type of settlement.

(1) *Uniform settlement.* A uniform settlement of all the foundations of a bridge will have little effect upon the structure. Settlements of nearly 1 foot have been experienced by small (70-foot), single-span bridges with no sign of appreciable distress.

(2) *Differential settlement.* Differential settlement can produce serious distress in any bridge. Where the differential settlement occurs between different substructure units, the magnitude of the damage depends on the bridge type and span length. Should a differential settlement take place beneath the footings of the same substructure, damage can vary from an opening of the vertical expansion joints between the wing wall and the abutment to severe tipping and cracking of walls or other members (See figures 5-19 through 5-21). Scour can cause support settlement (figure 5-22) or complete failure (figure 5-23).

c. Type of structure.

(1) *Simple (determinant).* As mentioned, the strength of a simple, or determinant, structure usually is not affected by movements unless they are quite large. There are usually enough joints to permit the movements without major damage to the basic integrity of the structure. At most, some finger joints or bearing may require resetting, or

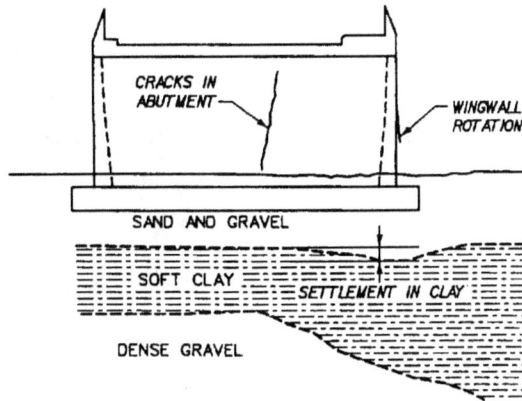

Figure 5-19. Differential settlement under an abutment.

Figure 5-20. Differential settlement.

beam supports may need shimming. However, pile bent or trestle bridges are very vulnerable since a

large settlement or movement of a bent could cause the superstructure to fall off a narrow bridge seat, leading to the loss of the bridge spans.

(2) *Indeterminant.* An indeterminant bridge is seriously affected by differential movements, since such movements at supports will redistribute the loads, possibly causing large overstresses. For example, a fixed-end arch could be severely damaged if a foundation rotates. Most continuous bridges have fewer joints than simple-span bridges. These bridges are very likely to be damaged if subjected to displacements which are greater in magnitude, or different in direction, from those that were considered in the original designs.

5-25. Indicators of movement

Foundation movements may often be detected by first looking for deviations from the proper geometry of the bridge. With the exception of curved structures, haunched members, and steeply inclined bridges, members and lines should usually be parallel or perpendicular to each other. While not always practical, especially for bridges spanning large bodies of water or for those located in urban industrial areas, careful observation of the overall structure for lines that seem incongruous with the rest of the bridge is a good starting point. For a more detailed inspection, the following methods are often useful:

a. *Check the alignment.* Any abrupt change or kink in the alignment of the bridge may indicate a lateral movement of a pier or of bearings. Older bridges are particularly vulnerable to ice pressures which can cause structural misalignment.

Figure 5-21. Differential pier movement causing superstructure movement.

Figure 5-22. Movement due to scour.

b. Sight along railings. A sudden dip in the rail line is often the result of settlement of a pier or abutment.

c. Run profile levels along the centerline and/or the gutter lines. This inspection technique will not only help to establish the existence of any settlement but will also identify any differential settle-

ments across the roadway. Normally, this kind of inspection technique will be employed only for large bridges or where information concerning the extent and character of differential settlement movement is required.

d. Check piers, pile bents, and abutment faces for plumbness with a transit. This inspection method provides an excellent check for the simpler techniques of plumbness determination. An out-of-plumb pier in either direction usually signifies foundation movement; it may also indicate a superstructure displacement. For small bridges and for preliminary checks, the use of a plumb bob is an adequate means for determining plumbness.

e. Observe the inclination of expansion rockers or roller movements. Rocker inclinations inconsistent with seasonal weather conditions may be a sign of foundation or superstructure movement. Of course, this condition may also indicate that the expansion rockers were set improperly. Out-of-plumb hangers on cantilevered structures are another indication of foundation shifting.

f. Observe expansion joints at abutments and walls. Observe the expansion joints for signs of opening or rotating. These conditions may indicate the movement of subsurface soils or a bearing failure under one of the footings.

g. Check deck joints and finger dams. Abnormally large or small openings, elevation differential, or jamming of the finger dams can be caused by substructure movements. Soil movements under the approach fills are also frequent occurrences.

h. Observe slabs, walls, and members. Cracks, buckling, and other serious distortions should be

Figure 5-23. Abutment failure from scour.

noted. Bracing, as well as the main supporting sections, should be scrutinized for distortion.

i. Check backwalls and beam ends. Check the backwalls for cracking which may be caused by either abutment rotation, sliding, or pavement thrust. Check for beam ends which are bearing against the backwall. This condition is a sign of horizontal movement of the abutment.

j. Observe fill and excavation slopes. Slide scarps, fresh sloughs, and seepage are indications of past or imminent soil movement.

k. Scour. See "Waterways," section VIII of this chapter.

l. Unbalanced postconstruction embankment or fills. Embankments or fills should be checked for balance and positioning. Unbalanced embankments or fills can cause a variety of soil movements which may impair the structural integrity of the bridge.

m. Underwater investigation of all piling and pile bents. Underwater investigation of piling and pile bents should be undertaken periodically. Check all timber piles for insect attack and deterioration. Examine steel piles well below the water surface. Steel piles protected in the splash zone can rust between the concrete jacket and the mudline. Examine prestressed piles below water for cracking or splitting.

5-26. Causes of foundation movements

The following causes of foundation movements, except as specifically noted, can produce lateral and/or vertical movements depending on the characteristics of the loads or substructures:

a. Slope failure (embankment slides). These are shear failures manifested as lateral movements of hillsides, cut slopes, or embankments. Footing or embankment loads imposing shear stresses greater than the soil shear strength are common causes of slides (figure 5-24, part a).

b. Bearing failures. Bearing failures are settlements or rotations of footings due to a shear failure in the soil beneath (figure 5-24, part b). When bearing or slope failures take place on an older structure, it usually indicates a change in subsurface conditions. This may endanger the security of nearby structures and foundations.

c. Consolidation. Serious settlement can result from consolidation action in cohesive soils. Settlement of bridge foundations may be caused by changes in the groundwater conditions, the placement of additional embankments near the structure, or increases in the height of existing embankments.

d. Seepage. The flow of water from a point of higher head (elevation or pressure) through the soil to a point of lower head is seepage (figure 5-24, part c). Seepage develops a force which acts on the soil through which the water is passing. Seepage results in lateral movement of retaining walls by:

(1) An increase in weight (and lateral pressure) of the backfill because of full or partial saturation.

(2) A reduction in resistance provided by the soil in front of the structure.

e. Water table variations. Large cyclic variations in the elevation of the water table in loose granu-

A. SLIDE FAILURE.

B. BEARING FAILURE.

C. SEEPAGE FORCES AT AN ABUTMENT.

D. DRAG FORCES ON PILES.

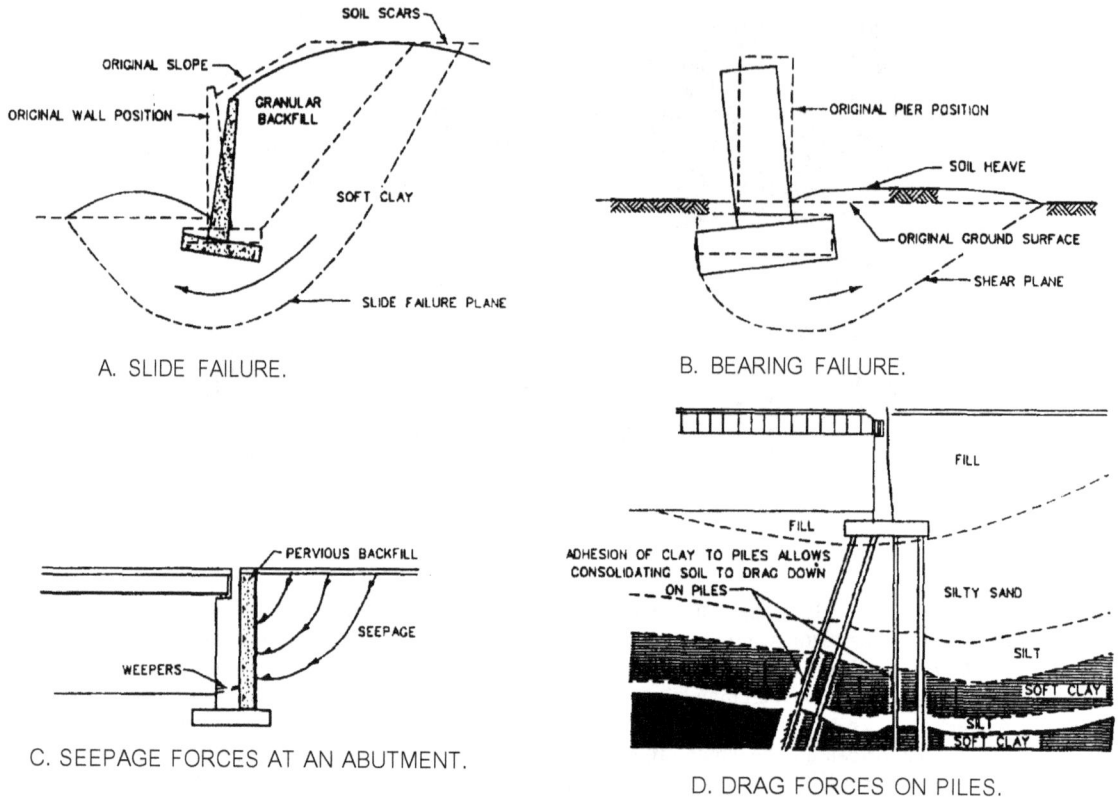

Figure 5-24. Causes of foundation movement.

lar soils may lead to a compaction of the upper strata. The effects of noncyclic changes in the water table such as consolidations, slides, and seepage were previously discussed. Changes in the water table may also change the characteristics of the soil which supports the foundation. Changes in soil characteristics may, in turn, result in the lateral movement or the settlement of the foundation.

f. Frost action. Frost heave in soil is caused by the growth of ice lenses between the soil particles. Footings located above the frost line may suffer from the effects of frost heave and a loss in bearing capacity due to the subsequent softening of the soil. The vertical elements on light trestle bents may also be lifted by frost and ice actions.

g. Expansive soils. Some clays, when wet, absorb water and expand, placing large horizontal pressures on any wall retaining such soil. Structures founded on expansive clay may also experience vertical soil movements (reverse settlement).

h. Ice. Ice can cause lateral movement in two ways. Where fine-grained backfill is used in retaining structures and the water table is above the frost line, the expansion of freezing water will exert a very large force against a wall. The piers

of river bridges are also subject to tremendous lateral loads when an ice jam occurs at the bridge.

i. Thermal forces from superstructures. On structures without expansion bearings, or where the expansion bearings fail to operate, thermal forces may tip the substructure units. Pavement thrust is another force that will have the same effect.

j. Drag forces. Additional embankment loads or very slow consolidation of a subsurface compressible stratum will exert vertical drag forces on the bearing piles which are driven through such material. This may cause yielding or failure of the piles (figure 5-24, part d).

k. Deterioration, insect attacks, and construction defects. All piles may develop weaknesses leading to foundation settlements from one or more of these causes:

(1) Timber, steel, and concrete piles are subject to loss of section because of decay, rusting, and deterioration.

(2) Timber piles are vulnerable to marine borers and ship worms.

(3) Construction defects include overdriven piles, underdriven piles, failure to fill pile shells

completely with concrete, or imperfect casings of a cast-in-place pile. Any of these defects will produce a weaker pile. Settlement will probably be gradual in improperly driven piles or in piles with weak or voided concrete. Piles suffering severe loss of section due to rust, spalling, chemical action, or insect infestation may fail suddenly under an unusually heavy load.

l. Scour and erosion. Scour can cause extensive settlement and/or structural failure as previously shown in figures 5-22 and 5-23. Since water will carry off particles of soil in suspension, a consider-

able hole can be formed around piers or other similar structural objects. This condition results in a greater turbulence of water and an increased size of soil particles that can be displaced. Scour is a very important consideration and will be given considerable attention in section VIII of this chapter. Erosion of embankments due to improper drainage (figure 5-25) can also lead to approach and abutment settlements.

m. Earth or rock embankments (stockpiles). Post-construction placement of embankments may cause instability since it will produce greater loads than were included in the original design.

Figure 5-25. Embankment erosion due to improper drainage. (Sheet 1 of 2)

Figure 5-25. Embankment erosion due to improper drainage. (Sheet 2 of 2)

Section VIII. WATERWAYS

5-27. General

A typical flow profile through a bridge is shown in figure 5-26. Note that the presence of a bridge (obstruction) does significantly affect the flow. Waterways should be inspected to determine whether any condition exists that could cause damage to the bridge or to the area surrounding the bridge. In addition to inspecting the channel's present condition, a record should be made of significant changes that have taken place in the channel, attributable to natural or artificial causes. When significant changes have occurred, an investigation must be made into the probable or potential effects on the bridge structure. Events which tend to produce local scour, channel degradation, or bank erosion are of primary importance.

5-28. Types of movement and effects on waterways

a. Scour. Scour is defined as the removal and transportation of material from the bed and banks of rivers and streams as a result of the erosive action of running water. Some general scouring (figure 5-27) takes place in all streambeds, particularly at flood stage. The characteristics of the channel influence the amount and nature of scour. Accelerated local scouring (figure 5-28) occurs where there is an interference with the streamflow, e.g., approach embankments extended in the river or piers and abutments constructed on the river bottom. The amount of scour in such cases depends on the degree to which streamflow is disturbed by the bridge and on the susceptibility of river bottom to scour action. Scour depth may range from zero in hard rock to 30 feet or more in very unstable river bottoms. In determining the depth of local scour, it is necessary to differentiate between true scour and apparent scour. As the water level subsides after flooding, the scour holes that are produced tend to refill with sediment. Elevations taken of the streambed at this time will not usually reveal true scour depth. However, since material borne and deposited by water will usually be somewhat different in character from the material in the substrate, it is often possible to determine the scour depth on this basis. If, for example, a strata of loose sand is found overlying a hard till substrate, it is reasonable to assume that the scour extends down to the depth of the till. This can often be confirmed by sounding or probing, provided the scour depth is limited to a few feet. Where coarse deposits or clays are encountered, sounding will probably be unsuccessful. Scour problems should be remedied as soon as possible since every flood can destroy the bridge totally. Typical situations which tend to lead to scour problems are as follows:

(1) *Sediment deposits.* The construction of an upstream dam, as seen in figure 5-29, will cause sediment previously carried downstream to be deposited in the reservoir, which acts as a settling basin. The increased scour capability of the downstream flow may degrade the lower channel.

(2) *Pier scour.* Scour around piers (figure 5-30) is greatly influenced by the shape of the pier and its skew to the direction of the flood flow. Note that the direction of flood flow will often be different from that of normal channel flow.

(3) *Loose riprap.* Loose riprap (figure 5-31) piled around piers to prevent local scour at the pier may cause deep scour holes to form downstream.

(4) *Lined banks.* Lined banks (figure 5-32) tend to reduce scour, but such a constriction might increase general scour in the bridge opening, especially at an adjacent or end pier.

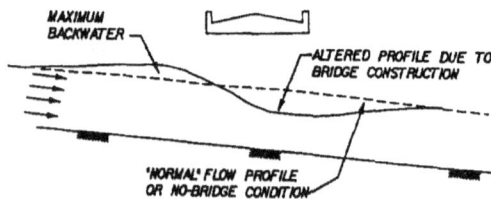

A. FLOW PROFILE

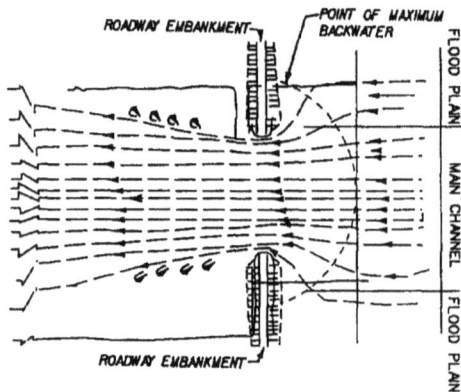

B. TYPICAL PLAN VIEW

Figure 5-26. Typical flow characteristics through a bridge.

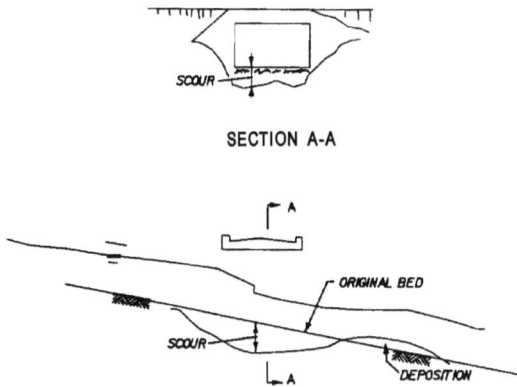

SECTION A-A

Figure 5-27. General scour

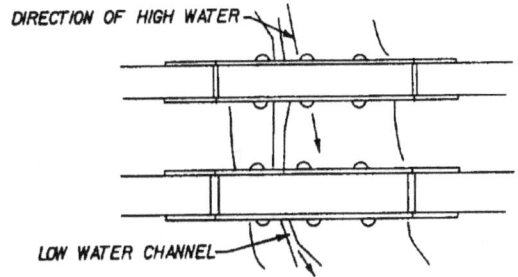

Figure 5-30. Pier scour.

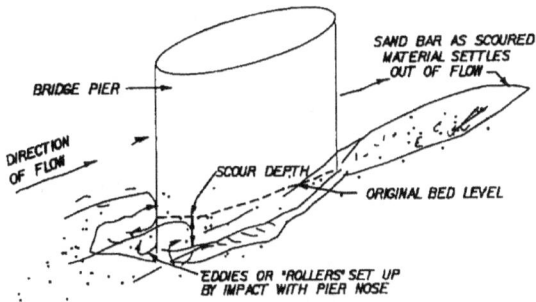

Figure 5-28. Localized scour.

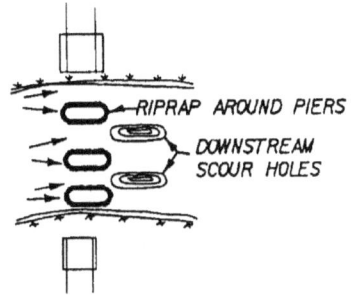

Figure 5-31. Loose riprap.

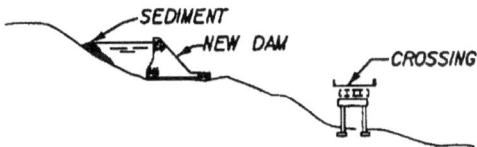

Figure 5-29. Sediment deposits.

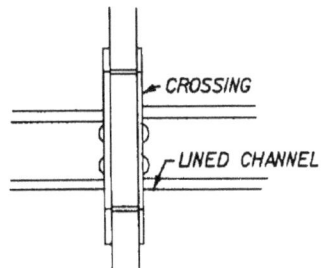

Figure 5-32. Lined banks.

(5) *Horizontal of vertical channel constrictions.* A firm or riprapped bottom or a horizontal constriction can cause a deep scour hole downstream with severe bank erosion resulting in downstream ponding as shown in figure 5-33.

(6) *Flooding.* During flood (figure 5-34) the waterway constriction may produce general scouring in the vicinity of the bridge.

(7) *Protruding abutments.* Protruding abutments (figure 5-35) may produce local scour. Deepest scour usually takes place at the upstream corner. The severity of scour increases with increased constriction.

(8) *Debris.* Collection of debris around piers (figure 5-36), in effect, enlarges the size of the pier and causes increased area and depth of scour.

(9) *River bends.* As seen in figure 5-37, a high scour potential exists for bridges located in the bend of the channel.

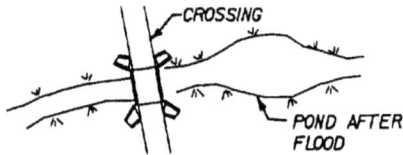

Figure 5-36. Debris problems.

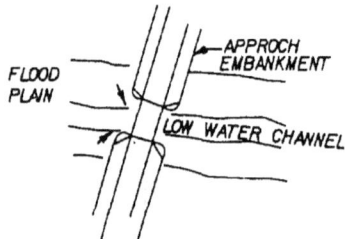

Figure 5-33. Channel constrictions.

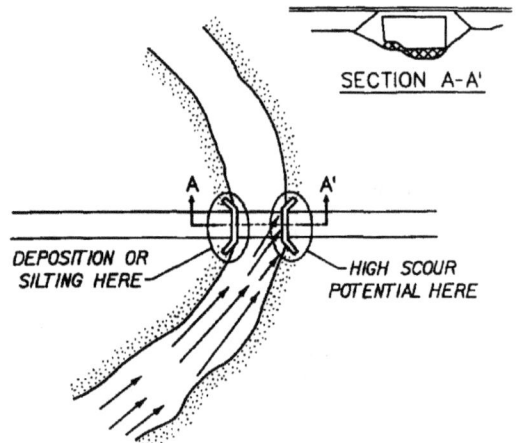

Figure 5-37. Bridge in a river bend.

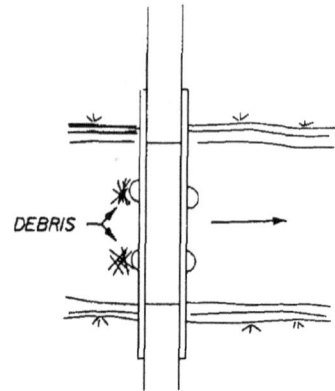

Figure 5-34. Flooding.

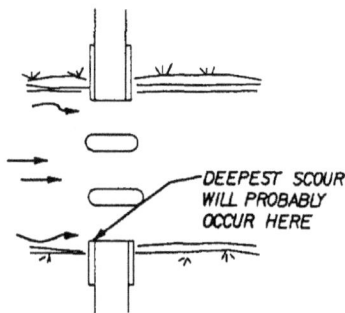

Figure 5-35. Protruding abutments.

b. Channel of streambed degradation. Streambed degradation is usually due to artificial or natural alteration in the width, alignment, or profile of the channel, upsetting the equilibrium or regime of the channel. These alterations may take place at the bridge site or some distance upstream or downstream. A channel is in regime if the rate of flow is such that it neither picks up material from the bed nor deposits it. In the course of years, the channel will gradually readjust itself to the changed condition and will tend to return to a regime condition. Streambed degradation and scour seriously endanger bridges with foundations located in erodible riverbed deposits and where the foundation does not extend to a depth below that of anticipated scour. Removal of material adjacent to the foundation may produce lateral slope instability causing damage to the bridge. Concrete slope protection (figure 5-38) or riprap (figure

5-39) is often provided to prevent bank erosion or to streamline the flow. It is particularly important where flow velocities are higher or where considerable turbulence is likely. It may also be necessary where there is a change in direction of the waterway. Slope cones around abutments are very susceptible to erosion and are usually protected. Situations which lead to channel degradation are as follows:

(1) *Channel change.* Changes in the channel (figure 5-40) steepen the channel profile and increase flow velocity. The entire upstream reach may degrade.

(2) *Removal of material.* Removal of large quantities of material (figure 5-41) (such as by dredging or gravel borrow pits) from the down-stream channel will cause increased upstream flow velocities and thus degradation.

(3) *Removal of obstruction.* A downstream obstruction (figure 5-42) will cause the flow under the bridge to be deep and slow. Once the obstruction is removed, the flow becomes more shallow and more rapid, causing degradation.

c. Waterway adequacy. Scour and streambed degradation are actually the result of inadequate waterway areas (freeboard). The geometry of the channel, the amount of debris carried during high water periods, and the adequacy of freeboard should be considered in determining waterway adequacy. Where large quantities of debris and ice are expected, sufficient freeboard is of the greatest importance.

Figure 5-38. Concrete slope protection.

Figure 5-39. Riprap slope protection.

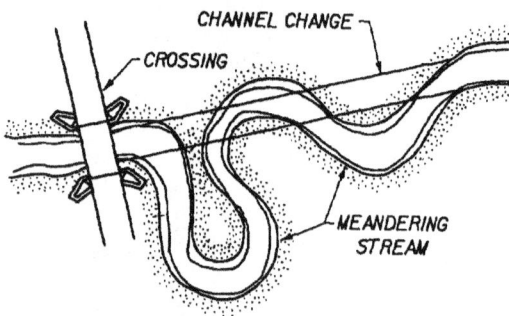

Figure 5-40. Channel change.

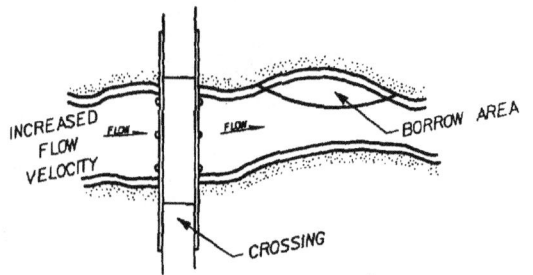

Figure 5-41. Material removal.

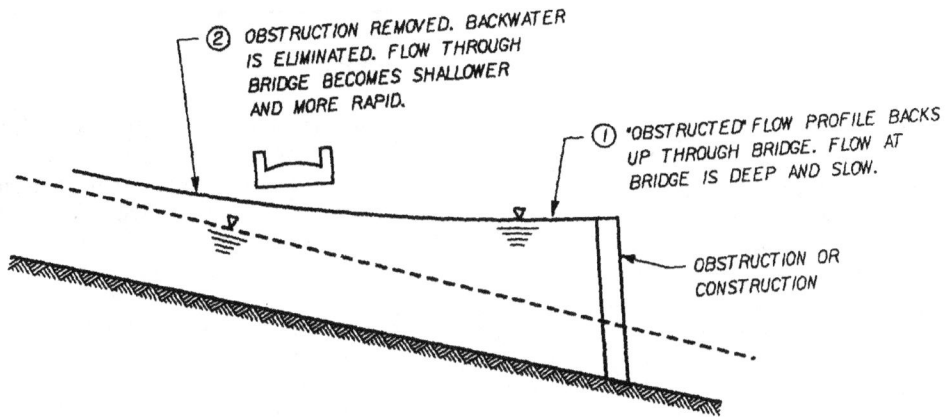

Figure 5-42. Obstruction removal.

CHAPTER 6

BRIDGE REDUNDANCY AND FRACTURE CRITICAL MEMBERS (FCMs)

Section I. GENERAL

6-1. Introduction

Due to the nature of their construction and their usage within the structure, some bridge members, referred to as fracture critical members or FCMs, are more critical to the overall safety of the bridge and, thus, are more important from the inspection standpoint. Although their inspection is more critical than other members, the actual inspection procedures for FCMs are no different. Therefore, the inspection of FCMs is not addressed separately in this manual and has been integrated into the normal inspection procedures discussed in this manual. The purpose of this chapter is to introduce the concept of FCMs and to provide guidelines for their identification.

6-2. Fracture critical members

The AASHTO manual, "Inspection of Fracture Critical Bridge Members," states that "Members or member components (FCMs) are tension members or tension components of members whose failure would be expected to result in collapse of the bridge." To qualify as an FCM, the member or components of the member must be in tension and there must not be any other member or system of members which will serve the functions of the member in question should it fail. The alternate systems or members represent redundancy.

6-3. Redundancy

With respect to bridge structures, redundancy means that should a member or element fail, the load previously carried by the failed member will be redistributed to other members or elements which have capacity to temporarily carry additional load, and collapse of the structure may be avoided. Redundancy in this manual is divided into three parts as further described:

a. *Load path redundancy.* Load path redundancy refers to the number of supporting elements, usually parallel, such as girders or trusses. For a structure to be nonredundant, it must have two or less load paths (i.e., load carrying members), like the bridges in figure 6-1 which only have two beams or girders. Failure of one girder will usually result in the collapse of the span, hence these girders are considered to be nonredundant and fracture critical. Examples of multiple load path

structures are shown in figure 6-2. There would be no FCMs in these structures.

b. *Structural redundancy.* Structural redundancy is defined as that redundancy which exists as a result of the continuity within the load path. Any statistically indeterminant structure may be said to be redundant. For example, a continuous two-span bridge has structural redundancy. In the interest of conservatism, AASHTO chooses to neglect structural redundancy and classify all two-girder bridges as nonredundant. The current viewpoint of bridge experts is to accept continuous spans as redundant except for the end spans, where the development of a fracture would cause two hinges which might be unstable.

c. *Internal redundancy.* With internal redundancy, the failure of one element will not result in the failure of the other elements of the member. The key difference between members which have internal redundancy and those which do not is the potential for movement between the elements. Plate girders, such as the one shown in figure 3-8, which are fabricated by riveting or bolting, have internal redundancy because the plates and shapes are independent elements. Cracks which develop in one element do not spread to other elements. Conversely, plate girders fabricated by rolling or welding, as shown in figures 3-7 and 3-8, are not internally redundant and once a crack starts to propagate, it may pass from piece to piece with no distinction unless the steel has sufficient toughness to arrest the crack. Internal redundancy is not ordinarily considered in determining whether a member is fracture critical but may be considered as affecting the degree of criticality.

6-4. Criticality of FCMs

The guidelines discussed above should be used to identify bridges that warrant special attention due to the existence of fracture critical members. Once an FCM is identified in a given structure, the information should become a part of the permanent record file on that structure. Its condition should be noted and documented on every subsequent inspection. The criticality of the FCM should also be determined to fully understand the degree of inspection required for the member. Criticality will be best determined by an experi-

Figure 6-1. Nonload path redundant bridges.

Figure 6-2. Load path redundant bridges.

enced structural engineer and should be based upon the following criteria:

a. *Degree of redundancy.* This was previously discussed in Section I.

b. *Live load member stress.* The range of live load stress in fracture critical members influences the formation of cracks. Fatigue is more likely when the live load stress range is a large portion of the total stress on the member.

c. *Propensity for cracking or fracture.* The fracture toughness is a measure of the material's resistance to crack extension and can be defined as the ability to carry load and to absorb energy in the presence of a crack. FCMs designed since 1978 by AASHTO standards are made of steel meeting minimum toughness requirements. On older bridges, coupon tests may be used to provide this information. If testing is not feasible, the age of the structure can be used to estimate the steel type which will indicate a general level of steel toughness. Welding, overheating, overstress, or member distortion resulting from collision may

adversely affect the toughness of the steel. FCMs that are known or suspected to have been damaged should receive a high priority during the inspection, and more sophisticated testing may be warranted.

d. *Condition of the FCMs.* A bridge that receives proper maintenance normally requires less time to inspect. Bridges with FCMs in poor condition should be inspected at more frequent intervals than those in good condition.

e. *Fatigue prone design details.* Certain design details have been more susceptible to fatigue cracking. Table 6-1 and figure 6-3 classify the types of details by category. The thoroughness of a fracture critical member inspection should be in the order of their susceptibility to fatigue crack propagation, namely from the highest (E) to the lowest (A) alphabetical classification.

f. *Previous and predicted loadings.* Repeated heavy loading is a consideration in determining the appropriate level of inspection. While this is not an exact science and new bridges have devel-

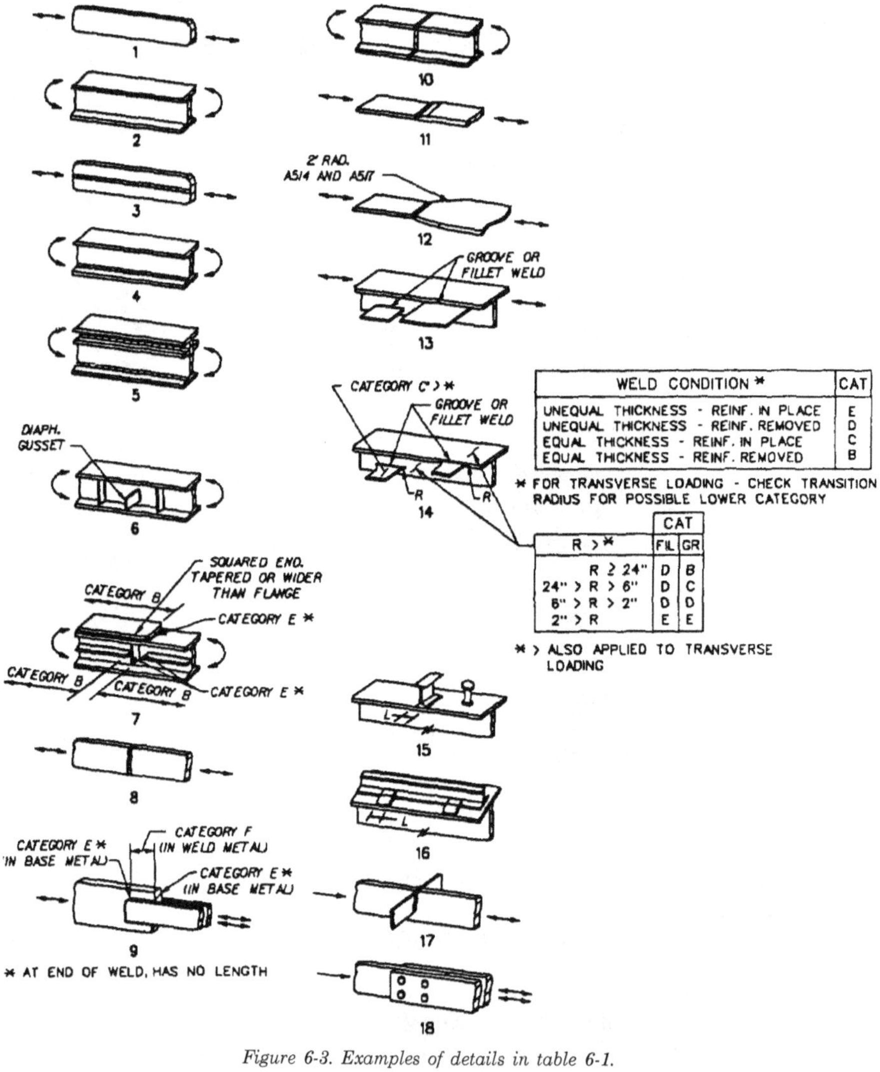
1
2
3
4
5
DIAPH. GUSSET
6
SQUARED END. TAPERED OR WIDER THAN FLANGE
CATEGORY B
CATEGORY E *
CATEGORY B
CATEGORY B
CATEGORY E *
7
8
CATEGORY F (IN WELD METAL)
CATEGORY E * (IN BASE METAL)
CATEGORY E * (IN BASE METAL)
9
* AT END OF WELD, HAS NO LENGTH
10
11
2' RAD. A514 AND A517
12
GROOVE OR FILLET WELD
13
CATEGORY C' > *
GROOVE OR FILLET WELD
R
R
14
WELD CONDITION * | CAT
UNEQUAL THICKNESS - REINF. IN PLACE | E
UNEQUAL THICKNESS - REINF. REMOVED | D
EQUAL THICKNESS - REINF. IN PLACE | C
EQUAL THICKNESS - REINF. REMOVED | B
* FOR TRANSVERSE LOADING - CHECK TRANSITION RADIUS FOR POSSIBLE LOWER CATEGORY
R >* | CAT FIL | GR
R ≥ 24" | D | B
24" > R > 6" | D | C
6" > R > 2" | D | D
2" > R | E | E
* > ALSO APPLIED TO TRANSVERSE LOADING
L
15
L
16
17
18

Figure 6-3. Examples of details in table 6-1.

oped fatigue cracks, the longer the bridge has been in service with a high volume of heavy loads, the greater the risk. When the precise number of loads experienced is not available, the location and the age is normally sufficient information to enable someone familiar with traffic in the area to make a reasonable estimate.

Section II. EXAMPLES

6-5. Two-girder system (or single-box girder)

a. Simple spans. A two-girder framing system is shown in figure 6-1. It is composed of two longitudinal girders which span between piers with transverse floorbeams between the girders. Floorbeams support longitudinal stringers. The failure of one girder may cause the span to collapse. These girders may be welded or riveted plate girders or steel boxbeams. The fracture critical elements in all of these girders are in the bottom flange and the web adjacent to the bottom flange as shown in figure 6-4, part a.

b. Anchor-cantilever. An anchor-cantilever span arrangement induces tension in the top flange and adjacent portion of the web in the area over the support as shown in figure 6-4, part b.

Table 6-1. Classification of types of details

General Condition	Type of Detail	Stress Category	Illustrative Example (See Figure 6-3)
Plain material	Base metal with rolled or cleaned surfaces. Flame cut edges with ASA smoothness of 1,000 or less.	A	1 , 2
Built-up members	Base metal and weld metal in members without attachments, built-up plates, or shapes connected by continuous full or partial penetration groove welds or by continuous fillet welds parallel to the direction of applied stress.	B	3,4,5,7
	Calculated flexural stress at toe of transverse stiffener welds on girder webs or flanges.	C	6
	Base metal at of partial length welded cover plates having square or tapered ends, with or without welds across the ends. (a) Flange thickness < 0.8 in. (b) Flange thickness > 0.8 in.	E E'	7 7
Groove welds	Base metal and weld metal, at full-penetration groove welded splices of rolled and welded sections having similar profiles when welds are ground flush, and weld soundness established by nondestructive inspection.	B	8,10,14
	Base metal and weld metal, in or adjacent to full-penetration groove welded splices at transitions in width or thickness, with welds ground to provide slopes no steeper than 1 to 2 1/2, with grinding in the direction of applied stress, and weld soundness established by nondestructive inspection.	B	11 , 12
	Base metal and weld metal, in or adjacent to full-penetration groove welded splices, with or without transitions having slopes no greater than 1 to 2 1/2 when reinforcement is not removed and weld soundness is established by nondestructive inspection.	C	8,10,11 12,14
	Base metal at details attached by groove welds subject to longitudinal loading when the detail length, L, parallel to the line of stress is between 2 in. and 12 times the plate thickness but less than 4 in.	D	13
	Base metal at details attached by groove welds subject to longitudinal loading when the detail length, L, is greater than 12 times the plate thickness or greater than 4 in. long.	E	13

c. Continuous spans. Continuous spans should be reviewed by a structural engineer or bridge designer to assess the actual redundancy and consequent presence of FCMs. In general, the fracture critical elements will be located near the center of the spans in the bottom of the girders and over the supports in the top of the girders.

6-6. Two-truss system

a. Simple spans. Most truss bridges have only two trusses. A truss may be considered a specialized girder with most of the web removed. Since tension members are the critical elements, the bottom chord and its connections are of primary concern. The diagonals and verticals which are in tension are also of primary concern. These members should be identified by a qualified structural engineer.

b. Anchor-cantilever. The anchor-cantilever in a truss system is similar to that in a girder system. In the area over an interior support (pier), the top chord is in tension. In the area near the end supports (abutments), the truss is similar to a simple-span truss and the same principles apply. From the center of the anchor span to the interior support, the stress arrangement is more complex and should be analyzed by a structural engineer.

c. Continuous spans. The statements regarding

Table 6-1. Classification of types of details-Continued

General Condition	Type of Detail	Stress Category	Illustrative Example (See Figure 6-3)
Groove welds (Cont'd)	Base metal at details attached by groove welds subjected to transverse and/or longitudinal loading regardless of detail length when weld soundness transverse to the direction of stress is established by nondestructive inspection. (a) When provided with transition radius equal to or greater than 24 in. and weld end ground smooth. (b) When provided with transition radius less than 24 in. but not less than 6 in. and weld end ground smooth. (c) When provided with transition radius less than 6 in. but not less than 2 in. and weld end ground smooth. (d) When provided with transition radius between 0 in. and 2 in.	B C D E	14 14 14 14
Fillet welded connections	Base metal at intermittent fillet welds.	E	--
	Base metal adjacent to fillet welded attachments with length L, in direction of stress less than 2 in. and stud-type shear connectors.	C	13,15, 16,17
	Base metal at details attached by fillet welds with detail lenth, L, in direction of stress between 2 in. and 12 times the plate thickness but less than 4 in.	D	13,15, 16
	Base metal at attachment details with detail length, L, in direction of stress (length of fillet weld) greater than 12 times the plate thickness or greater than 4 in.	E	7,9, 13,16
	Base metal at details attached by fillet welds regardless of length in direction of stress (shear stress on the throat of fillet welds governed by stress category F). (a) When provided with transition radius equal to or greater than 2 in. and weld end ground smooth. (b) When provided with transition radius between 0 in. and 2 in.	D E	14 14
Mechanically fastened connections	Base metal at gross section of high-strength bolted slip resistant connections, except axially loaded joints which induce out-of-plane bending in connected material.	B	18
	Base metal at net section of high-strength bolted bearing-type connections.	B	18
	Base metal at net section of riveted connections.	D	18
Fillet welds	Shear stress on throat of fillet welds.	F	9

continuous girders are also true regarding continuous trusses. In a continuous truss, the number of members in tension varies with the loading. Consequently, the determination of which members are in tension and which are fracture critical should be made by a structural engineer.

6-7. Cross girders and pier caps

The tension portions of simply supported cross girders and steel pier caps, as shown in figure 6-5, are nonredundant. These members usually consist of I sections or box beams. Unlike floorbeams which support only a portion of the deck, cross girders and steel pier caps support the entire end reactions of two longitudinal spans.

6-8. Supports and suspended spans

a. An example of a pin and hanger is shown in figure 3-15. Pin and hanger assemblies are as redundant as the framing system in which they are used. Hangers in a two-girder framing system offer no redundancy while the same assemblies used in a multibeam system have a high degree of redundancy.

b. An alternate support to the pin and hanger assembly is shown in figure 3-14. Portions of this detail (the short cantilever projection from the girder to the right) are fracture critical because part of it is in tension, and its failure will cause collapse unless it is used in a redundant framing system.

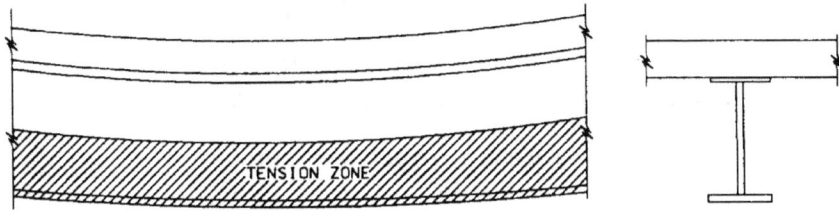

A. POSITIVE BENDING AT MIDSPAN

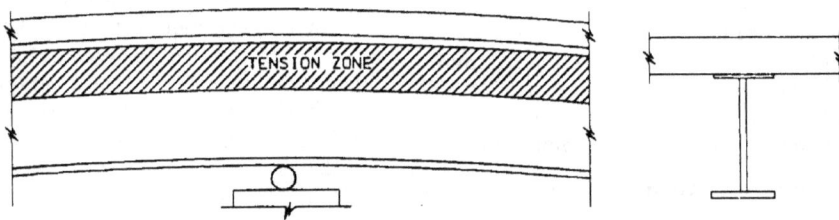

B. NEGATIVE BENDING OVER PIER

Figure 6-4. Portions of a girder in tension.

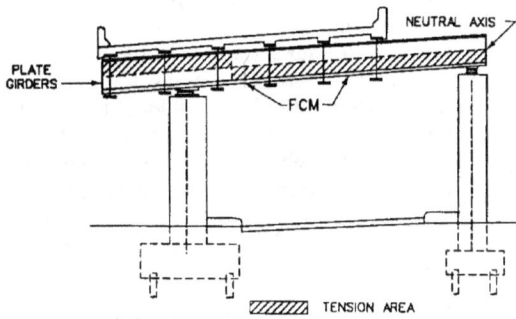

Figure 6-5. Steel cross girder on concrete piers.

CHAPTER 7

INSPECTION CONSIDERATIONS

Section I. TOOLS AND EQUIPMENT

7-1. Basic

For the inspection of bridges of any kind of material and structure, the bridge inspector should be equipped with at least a basic tool kit which includes, but is not necessarily limited to, the following: field books, inspection guide, sketch pad, paper, pencil, clipboard, keel marker, inspection mirror on a swivel head and extension arm for viewing difficult areas, camera (35 mm/Polaroid) for recording observed defects, safety belt for individual protection, tool belt, flashlight for viewing darkened areas, pocket knife, and binoculars.

7-2. Concrete inspection

In addition to the basic tools listed above, the concrete inspection tool kit should also consist of the following: 100-foot tape for measuring long cracks and large areas, 6-foot folding rule with 6-inch extender having 1/32-inch marking for measuring crack lengths and widths, piano wire for measuring the depth of cracks, chipping hammer for sounding concrete and removing deteriorated concrete, whisk broom for removing debris, scraper for removing encrustations, wire brush for cleaning exposed reinforcements, calipers (inside and outside) or micrometer for measuring exposed reinforcing bars, and a tape recorder for recording narratives of deteriorated conditions.

7-3. Steel inspection

In addition to the basic tools listed above, the steel inspection tool kit should also consist of the following: 100-foot tape for measuring long, deformed sections; 6-foot folding rule with 6-inch extender for measuring sections and offsets of deformed members; chipping hammer for cleaning heavily corroded areas; scraper for removing deteriorated paint and light corrosion; center punch for marking the end of cracks; calipers-dialed (inside and outside) or micrometer for measuring loss of section in webs, flanges, etc.; feeler gauges for measuring crack width; dry film paint gauge for measuring paint film thickness; large screwdriver; heavy-duty pliers; open-end wrench; wire brush for removing corrosion products; corrosion meter; dye penetrate kit and wiping cloths for examining small cracks; magnifying glass for viewing suspected areas and small cracks along welds and

around connections; shovel for removing debris; ultrasonic testing device; testing hammer for checking connections; and a cold chisel for marking reference points.

7-4. Timber inspection

In addition to the basic tools listed above, the timber inspection tool kit should also consist of the following: industrial crayon; chipping hammer for determining areas of unsound timber; ice pick for prying and picking to determine the extent of unsound timber; knife for prying and picking to determine the extent of unsound timber; prying tool for prying around fittings to determine, tightness, deterioration between surfaces, and extent of timber defects (DO NOT use a screwdriver); increment borer for taking test borings to determine extent of internal damage; creosoted plugs for plugging the holes made in the timber with the increment borer; pocket tape for measuring around piles or other members; 6-foot folding rule with 6-inch extender having 1/32-inch markings for measuring deteriorated areas; scraper for cleaning incrustations off pilings; 100-foot tape for measuring distances from reference points; whisk broom for removing debris; straight edge to be used as a reference point from which to measure section loss; calipers (inside and outside) or micrometer for measuring loss of section; testing hammer; and a cold chisel.

7-5. Cast iron, wrought iron, and aluminum inspection

In addition to the basic tools listed above, the tool kit should also consist of the following: 100-foot tape for measuring long deformed sections, 6-foot folding rule with 6-inch extender for measuring sections and offsets of deformed members, chipping hammer for cleaning heavily corroded areas, scraper for removing deteriorated paint and light corrosion, center punch for marking the end of cracks, calipers (inside and outside) or micrometer for measuring loss of section, feeler gauges for measuring crack width, large screwdriver, heavy-duty pliers, open end wrench, pocket knife, wire brush for removing corrosion products, corrosion meter, dye penetrate kit and wiping cloths for examining small cracks, magnifying glass for viewing suspected areas and small cracks along welds and around connections, and a shovel for removing debris.

7-6. Special equipment

Other specialty items which may be required are: ladders; scaffolds (travelers or cabling); "snooper" or "cherry picker" (truck-mounted bucket on a hydraulically operated boom (or on a platform truck)); burning, drilling, and grinding equipment; sand or shot blasting equipment; boat or barge; diving equipment (scuba or hard hat); sounding equipment (lead lines or electronic depth finders);

transit, level, or other surveying equipment; television camera for underwater use on closed circuit television with video tape recorder; magnetic or electronic locator for rebars; helicopters; air jet equipment; air breathing apparatus; mechanical ventilation equipment (blowers and air pipes); preentry air test equipment (devices to test oxygen content and to detect noxious gasses); ultrasound equipment; radiographic equipment; magnetic particle equipment; and dye penetrants. Ultrasound, radiographic, magnetic particle, dye penetrant, and other nondestructive methods are beyond the scope of the average inspection.

Section II. SAFETY

7-7. General

The safety of the bridge inspector is of the utmost importance. While the work may be hazardous, the accident probability may be limited by proceeding cautiously. Always be careful and use good judgment and prudence in conducting your activities both for your own safety and that of others. The safety rules must be practiced at all times to be effective.

7-8. Bridge site organization

a. *General.* The safety of personnel and equipment and the efficiency of the bridge inspection operation depend upon proper site organization.

b. *Personnel.* All individuals who are assigned to work aloft should be thoroughly trained in the rigging and use of their equipment, i.e., scaffolds, working platforms, ladders, and safety belts. The plans for inspection should give first consideration to safeguarding personnel from possible injuries.

c. *Equipment.* Inspection equipment, highway traffic barricades, and signs should be arranged according to the plan of inspection to accomplish as little handling, unnecessary movement, and prepositioning as possible. Vehicles not directly involved in the inspection process should be parked to prevent congestion and avoid interference in the areas of inspection.

d. *Orderliness.* Individuals should develop orderly habits in working and housekeeping on the job.

7-9. Personal protection

a. It is important to dress properly. Keep clothing and shoes free of grease. The following protective equipment is recommended for use at all times:

(1) Hard hat with a chin strap.

(2) Goggles, face masks, shields, or helmets, when around shot blasting, cutting, welding, etc.

(3) Reflective vests or belts when working in traffic.

(4) Life preservers or work vests when working over water.

(5) Shoes with cork, rubber, or some other nonslip soles.

b. If you wear glasses, wear them when climbing. The wearing of bifocals is an exception to this rule. Only regular single-lens glasses should be worn while climbing. Where work requiring close-up viewing is to be performed, a separate pair of glasses with lenses ground for this purpose should be worn.

c. Do not drink alcoholic beverages before or during working hours. They impair judgment, reflexes, and coordination.

7-10. Special safety equipment

a. A life line or belt must be worn when working at heights over 20 feet, above water, or above traffic.

b. A life-saving or safety skiff should be provided when working over large rivers or harbors; the skiff should have life preservers and life lines on board.

c. Warning signals, barricades, or flagmen are necessary when the deck is to be inspected or when scaffolding or platform trucks are used for access to the undersides or bridge seat of a structure.

7-11. Climbing of high steel

a. *General.* It is preferable to work from a traveler, catwalk, or platform truck, if possible. NOTE: On old ladders and catwalks, proceed with caution.

b. Scaffolding. When using scaffolding, the following precautions should be observed:

(1) Scaffolding and working platforms should be of ample strength and should be secure against slipping or overturning.

(2) Hanging scaffolds and other light scaffolds supported by ropes should be tested before using by hanging them 1 foot or so from the ground and loading them with a weight at least four times as great as their working load.

(3) Scaffolds should be inspected at least once each working day.

c. Ladders. Ladders should be used as working platforms only when it is absolutely necessary to do so. When using ladders, the following precautions should be observed:

(1) Make certain that the ladder to be used is soundly constructed. If made of wood, the material should be straight-grained and clear.

(2) Ladders should be tested to make sure they can carry the intended loads.

(3) Ladders should be blocked at the foot or tied at the top to prevent slipping.

(4) Personnel should be cautioned frequently about the danger of trying to reach too far from a single setting of a ladder.

d. Planks or platforms. Planks or platforms may be used where necessary. The following precautions should be observed:

(1) Planks should be large enough for the span.

(2) Never use a single plank. Two planks are a bare minimum; they should be attached by leads 18 inches apart.

e. Safety.

(1) Keep all catwalks, scaffolds, platforms, etc., free from ice, grease, or other slippery substances or materials.

(2) Catwalks, scaffolds, and platforms should have hand rails and toe boards to keep tools or other objects from being kicked off and become a hazard to anyone below.

(3) Always watch where you are stepping. Do not run or jump.

(4) Do not climb if you are tired or upset.

7-12. Confined spaces

a. General. In recent years there has been an increasing use of hollow structural members of the tubular or box-section types large enough to permit a man to enter the interior of the member. In bridge work, boxes are used both in large truss bridges and in girder bridges with rectangular or trapezoidal box sections. In these types of members, the interior is often closed off at both ends, forming a closed box. This protected interior is high in corrosion resistance even in the bare metal state. In some closed sections, a closable, water-tight, and vapor tight access hole is provided to permit inspection of the interior. While the box section offers both structural and maintenance advantages over other types of sections, there are certain health hazards of which the maintenance inspectors, and others who may be involved with these types of sections, should be aware.

b. Hazards. No health hazard exists in confined space if there is proper ventilation. However, a hazardous atmosphere can develop because of a lack of sufficient oxygen or because of a concentration of toxic gases. Oxygen deficiency can be caused by the low oxidation of organic matter which can become moistened. Toxic gasses may seep into the confined space or may be generated by such work processes as painting, burning, or welding. The confined space may be of such small volume that air contaminants are produced more quickly than the limited ventilation of the space can overcome. Persons should not be allowed to work in confined spaces containing less than 19 percent oxygen, unless provided with air breathing apparatus. However, as noted, a space with sufficient oxygen content can become unfit for human occupancy if the work conducted therein produces toxic fumes or gases. Such space should be occupied by the inspector only after adequate ventilation.

c. Safety procedures.

(1) *Preentry air tests.*

(a) Tests for oxygen content should be conducted with an approved oxygen-detecting device. A minimum of two tests should be conducted.

(b) Where the presence of other gases is suspected, tests for such gases should be conducted using approved gas-detecting devices. The following gases should be considered: carbon dioxide, carbon monoxide, hydrogen sulfide, methane, or any combustible gas.

(c) If the oxygen content of the air in the space is below 19 percent or if noxious gases present are equal or in excess of 125 percent of the Threshold Limit Values established by the American Conference of Government Industrial Hygienists (reference 6), no person should be allowed to enter such space until the oxygen content and gas content meet these specified limits for a minimum period of 15 minutes.

(2) *Ventilation during occupancy.*

(a) All confined spaces should be mechanically ventilated continuously during occupancy regardless of the presence of gas, the depletion of oxygen, or the conduct of contaminant-producing work.

(b) Where contaminant-producing operations are to be conducted, the ventilation scheme should be approved by an industrial hygiene engineer, safety engineer, marine chemist, or others qualified to approve such operations.

(3) *Air tests during occupancy.*

(a) If toxic gas presence or oxygen depletion is detected or suspected, air tests similar to the preentry air tests should be conducted during occupancy at 15-minute intervals.

(b) Where contaminant-producing operations are conducted, air tests should be conducted to determine the adequacy of the ventilation scheme.

Section III. DOCUMENTATION OF THE BRIDGE INSPECTION

7-13. General

The field inspection of a bridge should be conducted in a systematic and organized procedure that will be efficient and minimize the possibility of any bridge item being overlooked. Notes must be clear and detailed to the extent that they can be fully interpreted at a later date when a complete report is made. Sketches and photographs should be included in an effort to minimize long descriptions.

7-14. Planning and documenting the inspection

Careful planning of the inspection and selection of appropriate record keeping formats are essential for a well-organized, complete, and efficient inspection. During the planning phase the following items should be considered:

a. The inspection schedule.

b. The inspection type.

c. The resources required: manpower, equipment, materials, and special tools and instruments.

d. A study of all pertinent available information on the structure such as plans, previous inspections, current inventory report, and previous repairs.

e. *Optional documentation methods:*

(1) *Notebook.* The inspection notebook is normally used as the sole documentation on structures that are complex or unique. It should be prepared prior to the inspection and formatted to best facilitate the systematic inspection and recording of the bridge.

(2) *Bridge inspections.* For small and simple bridges, it may be more convenient to prepare checklists. Suggested items for these inspections are provided in appendixes B and C. Sketches can be drawn and additional comments provided as required. If available, standard prepared sketches should be attached with the coding of all members clearly indicated. Where sketches and narrative descriptions cannot fully describe the deficiency or defect, photographs should be taken and should be referred appropriately in the narrative. Prior to

the inspection, it should be determined which items are not applicable for the bridge to be inspected.

f. Coordination of resource requirements, particularly that of specialist personnel and special equipment.

g. The inspection procedure.

h. The existence of fracture critical members. As discussed in chapter 6, these members should be identified prior to the inspection so that they can be given special attention during the inspection. These members should be specifically denoted in the inspector's notebook prior to the inspection.

7-15. Structure evaluation

a. *General.* A bridge is typically divided into two main units, the substructure and the superstructure. For convenience the deck is sometimes considered as a separate unit. These basic units may be divided into structural members, which, in turn, may be further subdivided into elements or components. The general procedure for evaluating a structure is to assign a numerical rating to the condition of each element or component of the main units. A suggested numerical rating system is provided in appendix C. These ratings may be combined to obtain a numerical value for the overall condition of a member or of a unit.

b. *Explanatory aids.*

(1) *Narrative descriptions.* Descriptions of the condition should be as clear and concise as possible. Completeness, however, is essential. Therefore, narratives of moderate length will sometimes be required to adequately describe bridge conditions.

(2) *Photographs.* Photographs can be a great assistance. It is particularly recommended that pictures be taken of any problem areas that cannot be completely explained by a narrative description. It is better to take several photographs that may be unessential than to omit one that would preclude misinterpretation or misunderstanding of the report. At least two photographs of every structure should be taken. One of these should depict the structure from the roadway while the

other photo should be a view of the side elevation.

(3) *Summary.* An inspection is not complete until a narrative summary of the condition of the structure has been written.

(4) *Recommendations.* The inspector should list according to urgency any repairs that are necessary to maintain structural integrity and public safety.

Section IV. INSPECTION PROCEDURE

7-16. General

The development of a sequence for the inspection of a bridge is important since it actually outlines the plan for inspection. A well constructed sequence will provide a working guide for the inspector and ensure a systematic and thorough inspection.

a. Factors. Some of the factors that influence the procedure or sequence of a bridge inspection are:

(1) Size of the bridge.

(2) Complexity of the bridge.

(3) Existence of fracture critical members.

(4) Traffic density.

(5) Availability of special equipment.

(6) Availability of specialists.

b. Thoroughness of inspection. Thoroughness is as important as the sequence of the inspection. Particular attention should be given to:

(1) Structurally important members.

(2) Members most susceptible to deterioration or damage.

c. Visual inspection. Dirt and debris must be removed to permit visual observation and precise measurement. Careful visual inspection should be supplemented by appropriate special devices and techniques. If necessary, use of closed circuit television, photography, and mirrors will increase visual access to many components.

7-17. Inspection sequence

a. Average bridges. For bridges of average length and complexity, it is convenient to conduct the inspection in the following sequence:

(1) *Substructure units.*

(a) Piles.

(b) Fenders.

(c) Scour protection.

(d) Piers.

(e) Abutments.

(f) Skewbacks.

(g) Anchorages.

(h) Footings.

(2) *Superstructure units.*

(a) Main supporting members.

(b) Bearings.

(c) Secondary members and bracing.

(d) Utilities.

(e) Deck, including roadway and joints.

(f) End dams.

(g) Sidewalks and railings.

(3) *Miscellaneous.*

(a) Approaches.

(b) Lighting.

(c) Signing.

(d) Electrical.

(e) Barriers, gates, and other traffic control devices.

b. Large bridges. While the sequence of inspection for large bridges will generally be the same as for smaller bridges, exceptions may occur in the following situations:

(1) *Hazards.* Climbing and other hazardous tasks should be accomplished while the inspector is fully alert.

(2) *Weather.* Wind, extreme temperatures, rain, or snow may force the postponement of hazardous activities such as climbing, diving, or water-borne operations.

(3) *Traffic.* Median barriers, decks, deck joints, traffic control devices, and approaches should be inspected in daylight during periods of relatively light traffic to ensure inspector safety and to avoid the disruption of traffic.

(4) *Inspection party size.* When the inspection party is large, several different tasks may be performed simultaneously by different inspectors or groups of inspectors.

CHAPTER 8

BRIDGE COMPONENT INSPECTION

Section I. SUBSTRUCTURES

8-1. General

All bridge components are defined and discussed in chapter 3 of this manual. Bridge construction materials, their characteristics, and their associated deterioration problems are discussed in chapter 5 of this manual. This chapter presents a detailed, systematic guide to the inspection of each bridge component. Note that especially detailed instructions will be given for the inspection of FCMs. The guidelines provided in chapter 5 should be used as a supplement to this chapter to help recognize, describe, and assess problems with the various components.

8-2. Abutments

a. Check for scour or erosion around the abutment and for evidence of any movement (sliding, rotation, etc.) or settlement. Open cracks between adjoining wing walls or in the abutment stem, off-centered bearings, or inadequate or abnormal clearances between the back wall and the end beams are indications of probable movement (figure 8-1). If substructure cracking or movement is evidenced, a thorough subaqueous investigation or digging of test pits should be ordered to determine the cause of the problems.

b. Determine whether drains and weepholes are clear and functioning properly. Seepage of water through joints and cracks may indicate accumulation of water behind the abutment. Report any frozen or plugged weepholes. Mounds of earth immediately adjacent to weepers may indicate the presence of burrowing animals.

c. Check bearing seats for cracking and spalling, especially near the edges. This is particularly critical where concrete beams bear directly on the abutment. Check bearing seats for presence of debris and standing water.

d. Check for deterioration concrete in areas that are exposed to roadway drainage. This is especially important in areas where deicing chemicals are used.

e. Check backwalls for cracking and possible movement. Check particularly the construction joint between the backwall and the abutment.

f. Check stone masonry for mortar cracks, vegetation, water seepage, loose or missing stones, weathering, and spalled or split blocks.

8-3. Retaining walls

Inspection of most retaining walls should be similar to that of an abutment. Crib walls are subject to the same types of deterioration as other structures of wood, concrete, and steel:

a. Timber cribs may decay or be attacked by termites. However, the creosote treatment is usually very effective in protecting the wood.

b. Concrete cribs are subject to shipping and spalling. In addition, the locking keys or flanges at the ends of the crib pieces are sometimes broken off by vandals or inadvertently damaged by casual passersby.

c. Settlement of soil under the embankment will lead to distortion and possible damage to a crib wall. If sufficient movement occurs, the wall may fail.

8-4. Piers and bents

a. Check for erosion or undermining of the foundation by scour and for exposed piles (figure 8-2). Check for evidence of tilt or settlement as discussed in section VII of chapter 5. If problems of this type are evidenced, a thorough subaqueous investigation should be ordered to determine the cause of the problems.

b. Check for disintegration of the concrete, especially in the splash zone, at the waterline, at the groundline, and wherever concrete is exposed to roadway drainage (figure 8-3).

c. Check the pier columns and the pier caps for cracks.

d. Check the bearing seats for cracking and spalling.

e. Check stone masonry piers and bents for mortar cracks, water and vegetation in the cracks, and for spalled, split, loose, or missing stones.

f. Check steel piers and bents for corrosion (rust, especially at joints and splices). Bolt heads, rivet heads, and nuts are very vulnerable to rust, especially if located underwater or in the base of a column.

g. Examine grout pads and pedestals for cracks, spall, or deterioration.

h. Examine steel piles both in the splash zone and below water surface.

i. Investigate any significant changes in clearance for pier movement.

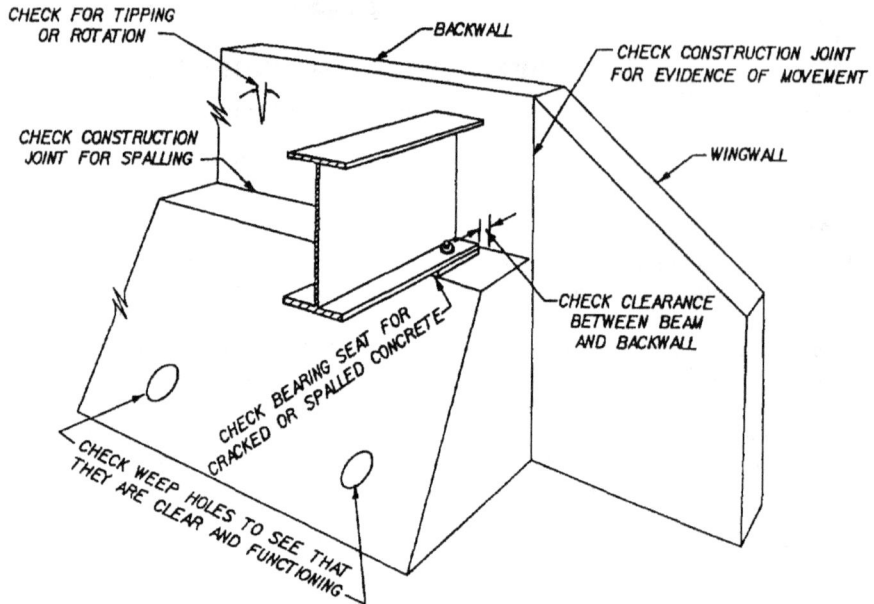

Figure 8-1. Abutment checklist items.

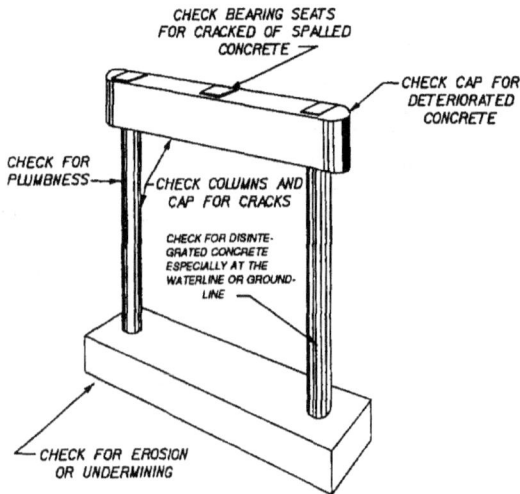

Figure 8-2. Concrete pier and bent checklist items.

Figure 8-3. Pier cap disintegration due to roadway drainage.

j. Check all pier and bent members for structural damage caused by collision or overstress.

k. Observe and determine whether unusual movement occurs in any of the bent members during passage of heavy loads.

l. Where rocker bents (figure 8-4) are designed to rotate freely on pins and bearings, check to see that such movement is not restrained. Restraint can be caused by severe corrosion or the presence of foreign particles.

Figure 8-4. Steel rocker bent.

m. Determine whether any earth or rock fills have been piled against piers causing loads not provided for in the original design and producing unstable conditions.

n. Inspect cross girder pier caps (figure 6-5). Their failure will generally cause collapse of the unsupported span. Therefore, they are considered FCMs and should be closely inspected as follows:

(1) *Riveted.*

(a) Check all rivets and bolts to determine that they are tight and that the individual components are operating as one. Check for cracked or missing bolts, rivets, and rivet heads.

(b) Check the member for misplaced holes or repaired holes that have been filled with weld metal. These are possible sources of fatigue cracking.

(c) Check the area around the floorbeam and lateral bracing connections for cracking in the web due to out-of-plane bending.

(d) Check the entire length of the tension flanges and web for cracking which may have originated from corrosion, pitting or section loss, or defects in fabrication (e.g., nicks and gouges in the steel).

(e) Check entire length for temporary erection welds, tack welds, or welded connections not shown on the design drawings.

(2) *Welded.*

(a) Check all tranverse groove welds for indication of cracks, especially near backup bars.

(b) Check all tranverse stiffeners and connection plates at the connection to the web, particularly at floorbeams and lateral bracing where out-of-plane bending is introduced.

(c) If longitudinal stiffeners have been used, check any butt weld splices in the longitudinal stiffeners. The web at the termination of longitudinal stiffeners should also be checked carefully.

(d) If cover plates are present, check carefully at the terminus of each for cracks.

(e) Observe any area of heavy corrosion for pitting section loss or crack formation.

(f) If girders have been haunched by use of insert plates, observe the transverse groove welding between the web and insert plate.

(g) Check longitudinal fillet welds for possible poor quality or irregularities that may cause cracking to initiate. This is especially important during the first inspection of the member so that defects can be recorded and properly documented on follow-up inspections.

(h) Check for cracks at any intersecting fillet welds. If triaxial intersecting welds are found on an FCM, they should be reported and carefully examined in future inspections.

(i) Check any plug welds.

(j) Check bolted splices for any sign of cracks in girders or splice plates and look for missing or cracked bolts.

(k) Check the entire length of the tension flanges and web for cracking which may have originated from corrosion, pitting or section loss, or defects in fabrication.

(l) Check entire length for temporary erection, tack welds, or welded connections not shown on the design drawings.

8-5. Pile bents

a. Concrete. Check for the same items as discussed in paragraph 8-4.

b. Timber.

(1) Check for decay in the piles, caps, and bracing (figure 8-5). The presence of decay may be determined by tapping with a hammer or by test boring the timber. Check particularly at the groundline, or waterline, and at joints and splices, since decay usually begins in these areas.

(2) Check splices and connections for tightness and for loose bolts.

(3) Check the condition of the cap at those points where the beams bear directly upon it and at those points where the cap bears directly upon the piles. Note particularly any splitting or crushing of the timber in these areas.

(4) Observe caps that are under heavy loads for excessive deflections.

(5) Check for rotted or damaged timbers in the backwalls of end bents (abutments), especially where such conditions would allow earth to spill upon the caps or stringers. Approach fill settlement at end bents may expose short sections of piling to additional corrosion or deterioration.

(6) In marine evironments, check structures for the presence of marine borers and shipworms.

(7) Check timber piles in salt water to determine damage caused by marine borers.

(8) Check timber footing piles in salt water exposed by scour below the mudline for damage caused by marine borers.

(9) Check timber piles in salt water at checks in the wood, bolt holes, daps, or other connections for damage by marine borers.

c. Steel.

(1) Check the pile bents for the presence of rust, especially at the ground level line. Use a chipping hammer, if necessary, to determine the extent of the rust. Over water crossings, check the splash zone (2 feet above high tide or mean water level) and the submerged part of the piles for indications of rust.

(2) Check for debris around the pile bases. Debris will retain moisture and promote rust.

(3) Check the steel caps for rotation due to eccentric (off-center) connections.

(4) Check the bracing for broken connections and loose rivets or bolts.

(5) Check condition of web stiffeners.

8-6. Dolphins and fenders

a. Steel. Observe the "splash zone" carefully for severe rusting and pitting. The splash zone is the area from high tide to 2 feet above high tide. Where there are no tides, it is the area from the mean water level to 2 feet above it. Rusting is much more severe here than at midtide elevations.

b. Concrete. Look for spalling and cracking of concrete, and rusting of reinforcing steel. Be alert for hour-glass shaping of piles at the waterline.

c. Timber. Observe the upper portions lying between the high water- and mudline for marine

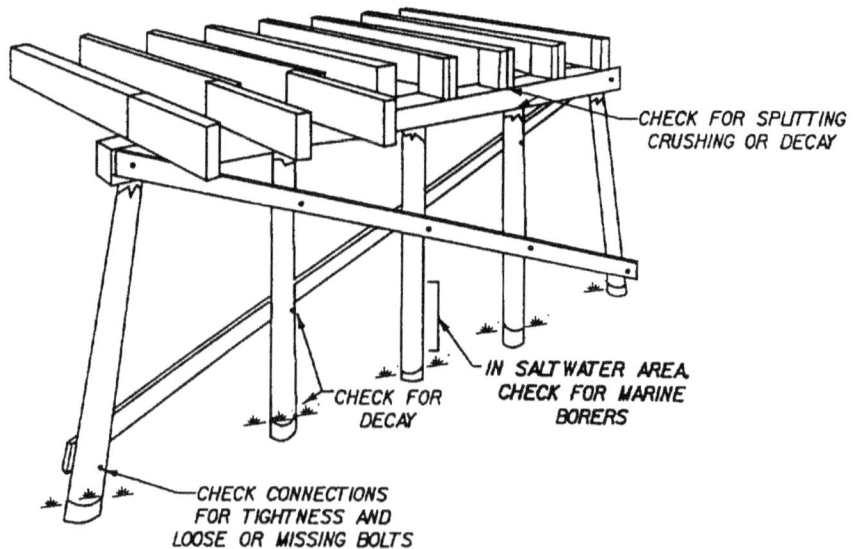

Figure 8-5. Timber bent checklist items.

insecs and decay (figure 8-6). Check the fender pieces exposed to collision forces for signs of wear.

d. Structural damage. Check all dolphins or fenders for cracks, buckled or broken members, and any other signs of structural failures or damage from marine traffic.

(1) Piling and walers require particular attention, since these are areas most likely to be damaged by impact.

(2) Note any loose or broken cable which would tend to destroy the effectiveness of the cluster (figure 8-6). Note whether they should be rewrapped.

(3) Note missing walers, blocks, and bolts.

e. Protective treatment. Note any protective treatment that needs patching or replacing. This includes breaks in the surface of treated timbers, cracks in protective concrete layers, rust holes or tears in metal shields, and bare areas where epoxy or coal tar preservatives have been applied externally.

f. Catwalks. Note the condition of the catwalks for fender systems.

Figure 8-6. Deteriorated timber dolphins.

Section II. SUPERSTRUCTURES

8-7. Concrete beams and girders

a. All beams.

(1) Check for spalling concrete, giving special attention to points of bearing where friction from thermal movement and high edge pressure may cause spalling (figure 8-7).

(2) Check for diagonal cracking, especially near the supports. The presence of diagonal cracks on the side of the beam may indicate incipient shear failure. This is particularly important on the older prestressed bridges. Cantilever bridges,

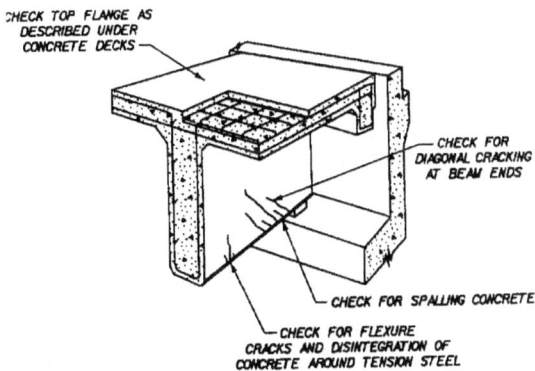

Figure 8-7. Concrete beam checklist.

whether of prestressed or reinforced concrete, utilize a shiplapped joint in which the suspended span rests upon bearings located on the anchor span (figure 8-8). The shiplap cantilevers with reentrant corners are fracture critical details and should be inspected very carefully for signs of cracking or other deterioration.

(3) Check for flexure (vertical) cracks or disintegration of the concrete, especially in the area of the tension steel. Discoloration of the concrete surface may be an indication of concrete deterioration or the corrosion of the reinforcing steel. In severe cases, the reinforcing steel may become exposed.

(4) Observe areas that are exposed to roadway drainage for disintegrating concrete.

(5) Check for damage caused by collision or fire.

(6) Note any excessive vibration or deflection during passage of traffic.

b. Box girders.

(1) Examine the inside of box girders for cracks and to see that the drains are open and functioning properly.

(2) Check the soffit of the lower slab and the outside face of the girders for excessive cracking.

(3) Check diaphragms for cracks.

Figure 8-8. Shiplapped cantilever joint.

(4) Examine the underside of the slab and top flange for scaling, spalling, and cracking.

(5) Note any offset at the hinges which might indicate problems with the hinge bearing. An abnormal offset should be investigated further to determine the cause and the severity of the condition.

c. *Prestressed concrete members.*

(1) Check for longitudinal cracks on all flange surfaces. This may occur on older prestressed bridges where insufficient stirrups were provided.

(2) Examine the alignment of prestressed beams.

(3) Check for cracking and spalling in the area around the bearings and at the cast-in-place diaphragms where differential creep and humping of the beams may have some ill effects.

(4) On pretensioned deck units, either box-beams or voided units, check the underside during the passage of traffic to see whether any unit is acting independently of the others.

8-8. Steel beams and girders

It should be remembered that steel beams and girders may qualify as FCMs (chapter 6). Any serious problems found in an FCM should be addressed immediately since its failure could cause total collapse of the bridge. Immediate closure of the bridge may be warranted if the defect is deemed serious. Regardless of the member's FCM status, the following items should be inspected:

a. Check members for cleanliness and freedom from debris, especially on the top side of the bottom flange. Unclean members should be espe-cially suspect since this indicates lack of maintenance and ideal conditions for deterioration. Cleaning may be necessary to properly inspect the members for cracks and corrosion.

b. Inspect steel for corrosion and deterioration (figure 8-9) especially at the following places:

(1) Along the upper flange.

(2) Around bolts and rivet heads.

(3) At gusset, diaphragm, and bracing connections.

(4) At cantilever hanger and pin connections.

(5) Under the deck joints and at any other points that may be exposed to roadway drainage.

(6) At any point where two plates are in face-to-face contact and water can enter (such as between a cover plate and a flange).

(7) At the fitted end of stiffeners.

(8) At the ends of beams where debris may have collected.

c. If rusting and deterioration are evident, check the members to determine the extent of reduced cross-sectional area, using calipers, rulers, corrosion meters, or section templates.

d. Check all rivets and bolts to determine that they are tight and that the individual components are operating as one. Check for cracked or missing bolts, rivets, and rivet heads.

e. Check the entire length of the tension flanges and web for cracking which may have originated from corrosion, pitting or section loss, or defects in fabrication (such as nicks and gouges in the steel).

f. Examine welds, weld terminations, and adjacent metal for cracks, particularly at:

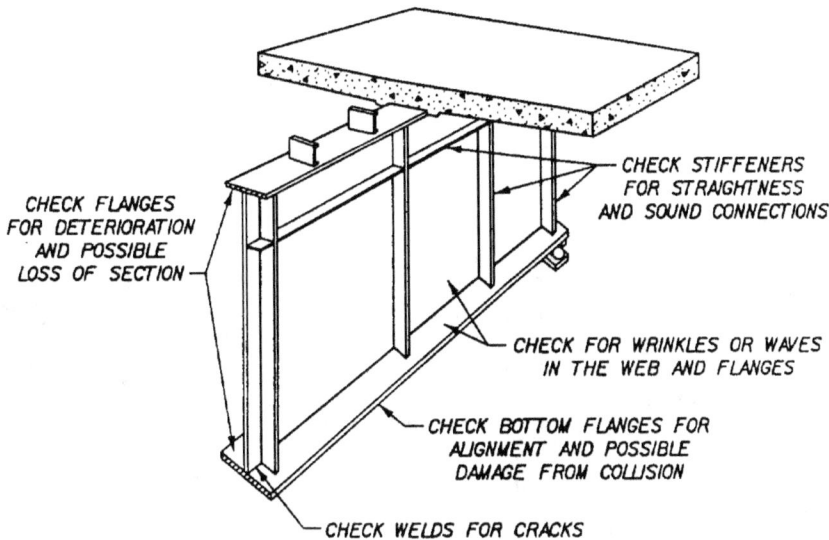

Figure 8-9. Steel girder checklist items.

(1) Unusual types of weld connections or connections to which access would have been difficult for the welder.

(2) Field welds, especially plug welds, often cause stress concentrations and are thus prone to fatigue cracking.

(3) Connections transmitting heavy torsional or in-plane moments to the members. Typical connections of this type are:

(a) Floorbeam-to-girder connections (see figure 8-10).

(b) Brackets cantilevered from the fascia beams (or any cantilever connection from a beam).

(c) Moment splices.

(d) Joints in rigid frames.

(4) Sudden changes in cross section or configuration or other locations subject to stress concentrations or fatigue loadings. Several specific areas in this category are:

(a) Termination points of welded cover plates (figure 8-11).

(b) Longitudinal welds along the length of cover plates, especially intermittent welds as shown in figure 8-12.

(c) Welds of insert plates in haunched girders (figure 8-13).

(5) The potential crack locations for longitudinal and transverse web stiffeners are summarized in figure 8-14. Note in figure 8-14 (sheet 3) that the crack is due to an improperly made butt weld.

(6) The intersection of horizontal and vertical fillet welds such as that shown in figure 8-15.

(7) Horizontal connection plates used to connect lateral bracing, as shown in figure 8-16.

(8) Areas where vibration and movement could produce fatigue stress.

(9) Coped sections/reentrant corners (figure 8-17).

(10) Connections of boxbeams to columns (figure 8-18).

g. Check the general alignment by sighting along the members. Misalignment or distortion may result from overstress, collision, or fire damage. If such a condition is present, its effect on structural safety of the bridge should be fully investigated.

h. Check for wrinkles, waves, cracks, or damage in the web flanges of steel beams, particularly near points of bearing. This condition may indicate overstressing. Check the stiffeners for straightness and determine whether their connections are broken, buckled, or pulled from the web.

i. Determine whether any unusual vibration or excessive deflections occur under the passage of heavy loads.

j. Check the wind locks (figure 3-14) for binding, jamming, improper fit, or excessive movement before engaging.

k. Thoroughly check the inside of box girders for all of these problems.

l. In composite construction, stud type shear connectors are utilized between the upper flange of the beam or girder and the deck slab. In this case, the underside of the top girder flange should be checked for cracks as shown in figure 8-19. This is

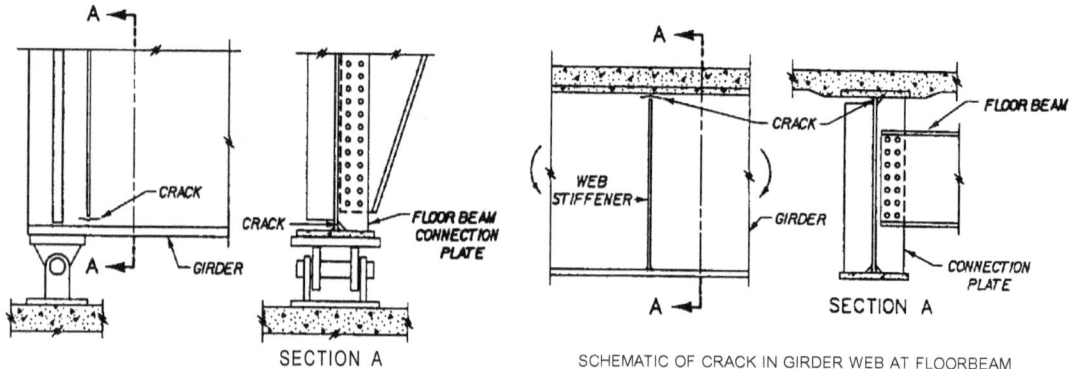

SCHEMATIC SHOWING CRACK IN GIRDER WEE AT
FLOOR BEAM CONNECTION PLATES AT SUPPORTS

SCHEMATIC OF CRACK IN GIRDER WEB AT FLOORBEAM
CONNECTION PLATES

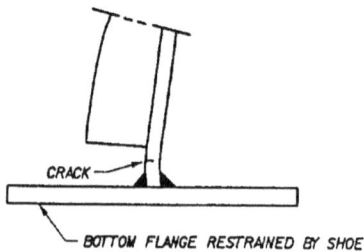

A. BOTTOM OF CONNECTION PLATE.

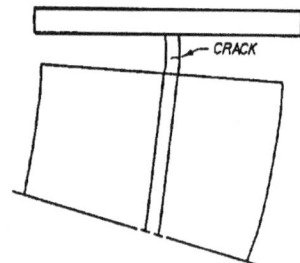

B. TOP OF CONNECTION PLATE.

Figure 8-10. Floorbeam connection plates.

only necessary in the areas spanning over pier supports; i.e., the negative moment region.

8-9. Pin and hanger connections

a. Hanger plates. These plates (figure 3-15) are as critical as the pin in a pin and hanger connection. It is, however, easier to inspect since it is exposed and readily accessible; and the following steps are required:

(1) Try to determine whether the hanger-pin connection is frozen, since this can induce large moments in the hanger plates. Check both sides of the plate for cracks due to bending of the plate from a frozen pin connection.

(2) Observe the amount of corrosion buildup between the webs of the girders and the back faces of the plates.

(3) Check the hanger plate for bowing or out-of-plane movement from the webs of the girders. If the plate is bowed, check carefully at the point of maximum bow for cracks which might be indicated by broken paint and corrosion.

(4) All welds should be checked for cracks.

b. Pin. Rarely is the pin directly exposed in a

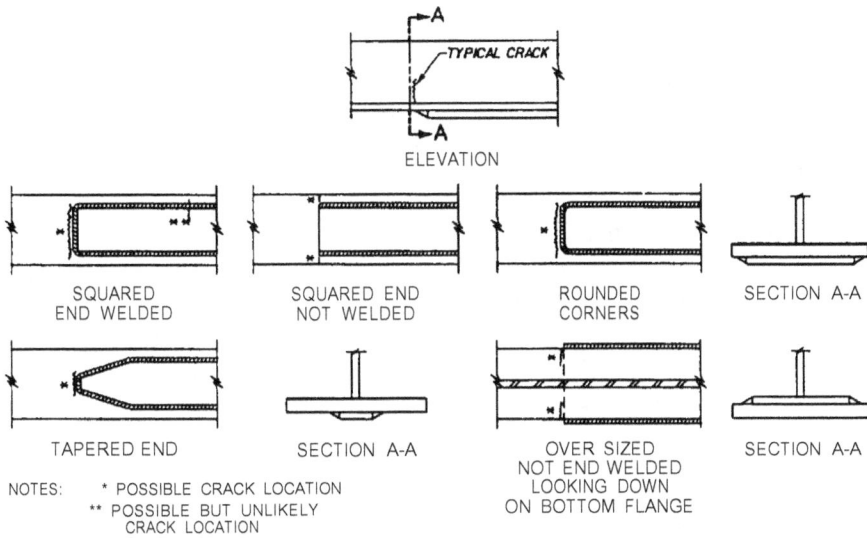

Figure 8-11. Cracks in ends of cover plates.

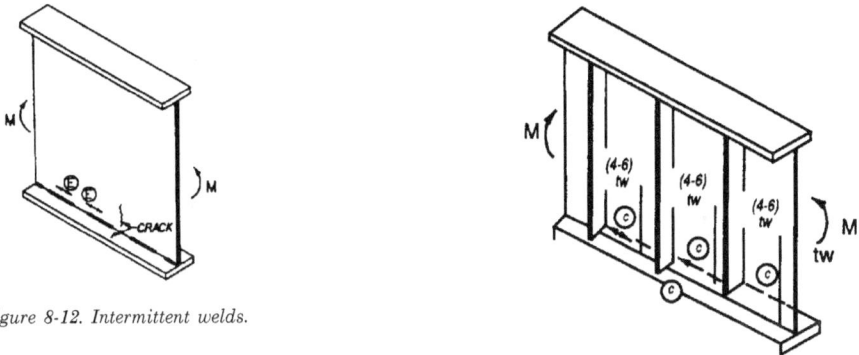

Figure 8-12. Intermittent welds.

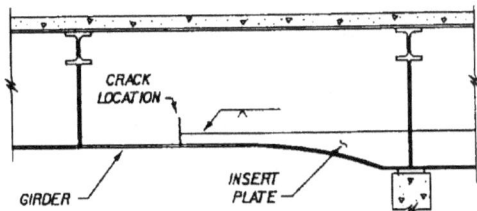

Figure 8-13. Insert plates in haunched girders.

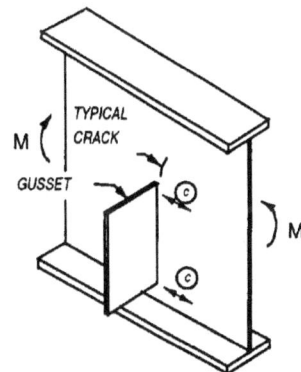

Figure 8-14. Attachments. (Sheet 1 of 3)

pin and hanger connection, and as a result its inspection is difficult but not impossible. By carefully taking certain measurements, the apparent wear can be determined. If more than 1/8-inch net section loss has occurred, it should be considered critical and given immediate attention. Several types of pins and hangers and the manner for

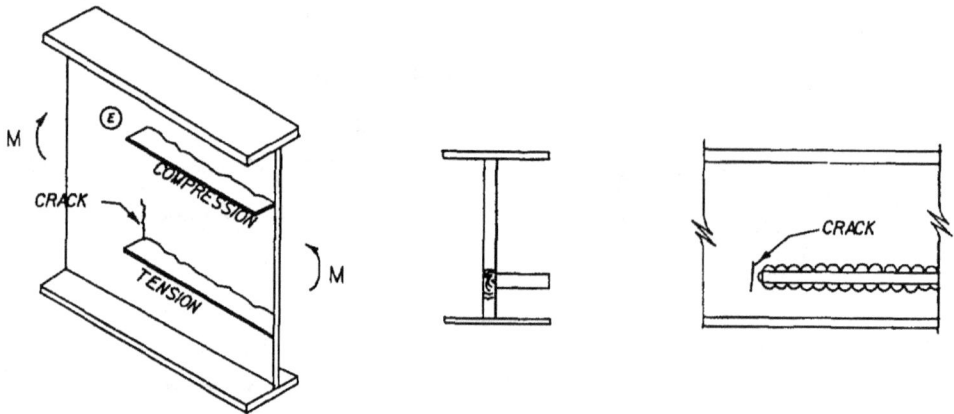

Figure 8-14. Attachments. (Sheet 2 of 3)

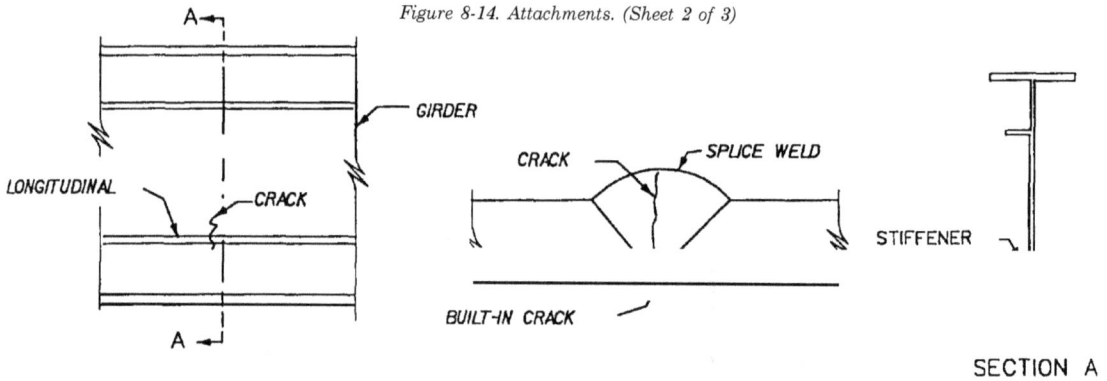

SCHEMATIC OF GIRDER SHOWING LONGITUDINAL STIFFENERS
IN TENSION STRESS REGION WIYH BUTT WELDED SPLICE
Figure 8-14. Attachments. (Sheet 3 of 3)

measuring wear on each are discussed in the following paragraphs:

(1) *Girder pin and hanger.* Wear to the pins and hangers will generally occur in two locations, at the top of the pin and top of the hanger on the cantilevered span, and the bottom of the pin and bottom of the hanger on the suspended span. Sometimes wear, loss of section, or lateral slippage may be indicated by misalignment of the deck expansion joints or surface over the hanger connection. The following inspection procedure should be used. Figure 3-16 can be used as a reference sketch.

(a) Locate the center of the pin.

(b) Measure the distance between the center of the pin and the end of the hanger.

(c) Compare to plan dimensions, if available. Remember to allow for any tolerances, since the pin was not machined to fit the hole exactly. Generally, this tolerance will be 1/32 inch. If plans are not available, compare to previous measurements. The reduction in this length will be the "apparent wear" on the pin.

(2) *Fixed pin and girder.* Wear will generally be on the top surface of the pin due to rotation from live load deflection and tractive forces. The following steps should be used with figure 3-16 as a reference:

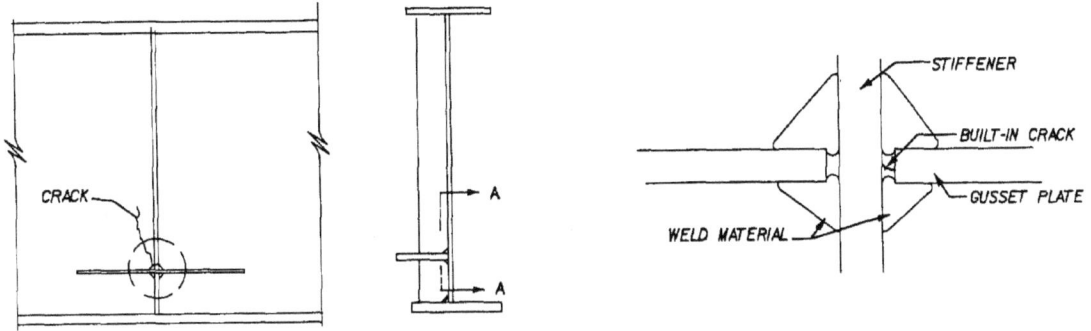

Figure 8-15. Intersecting welds.

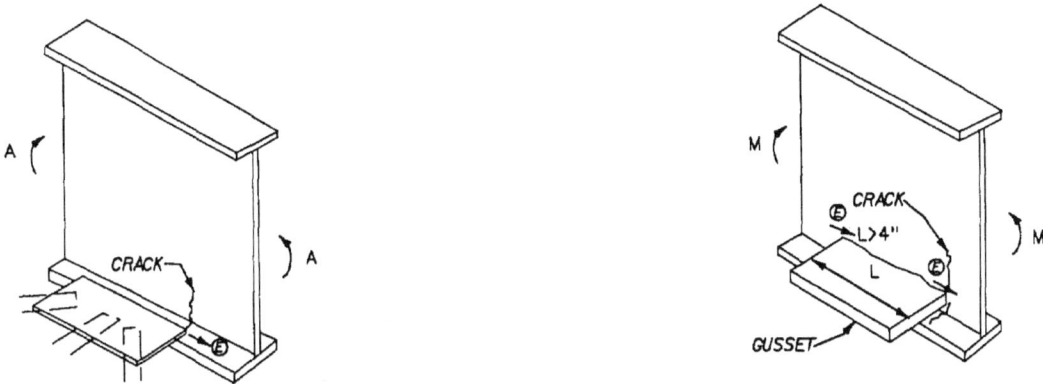

Figure 8-16. Flange and web attachments.

(a) Locate the center of the pin.

(b) Measure the distance between the center of the pin and some convenient fixed point, usually the bottom of the top flange.

(c) Compare this distance to the plan dimensions and determine the amount of section loss.

(3) *Truss pin and hunger*. Pin and hanger arrangements are slightly different when used in trusses. Usually the hanger plates are compact members similar to a vertical or diagonal. The hanger then slips between gusset plates at both the upper and lower chords. It is more difficult to find a fixed reference point because the gusset plate dimensions are not usually given on design plans, but two recommended options are the intersection of the upper or lower chord and the nearest diagonal or the edge of the gusset plate along the axis of the hanger. Both these points will provide readily identifiable reference points which can be recreated easily by the next inspection team. For this reason, measurements should be carefully documented along with the temperature and weather conditions. The inspection procedure should include:

(a) Locate center of pin.

a. COPED SECTION IN FLANGES

b. ACCEPTABLE AND UNACCEPTABLE COPES.

GUSSET

c. CUTOUTS IN GUSSET PLATES

d. BEAM TO GIRDER CONNECTIONS.

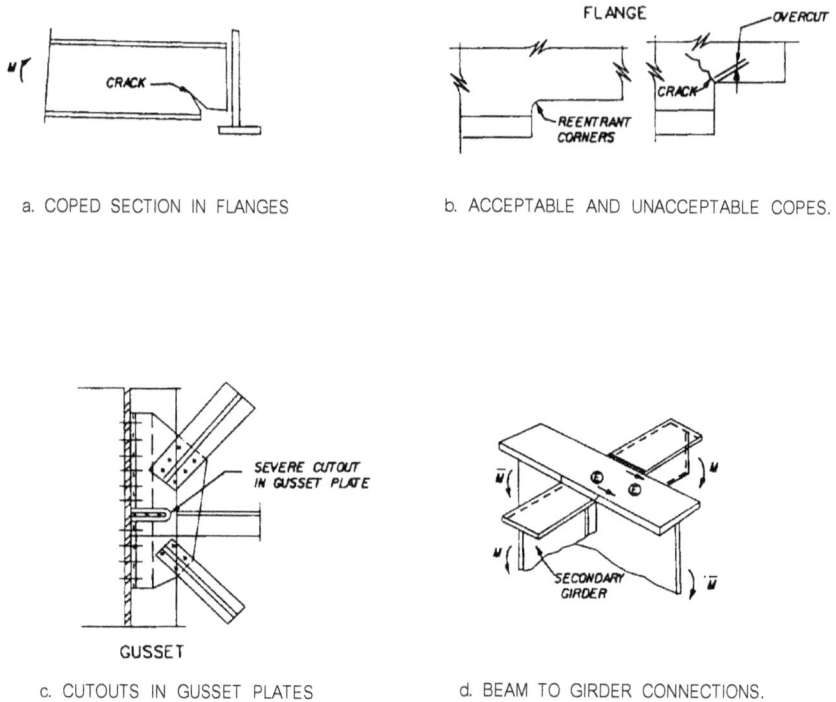

Figure 8-17. Copes and reentrant corners.

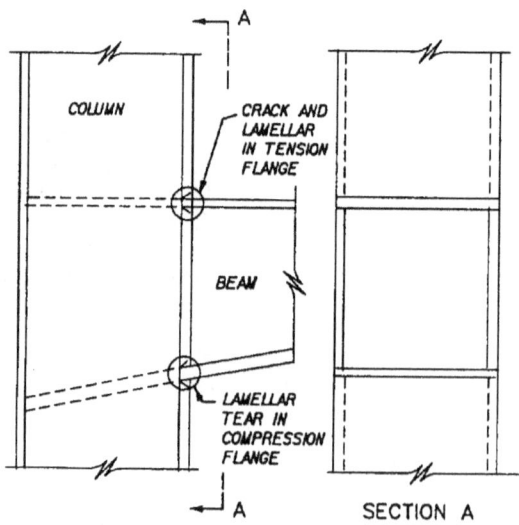

Figure 8-18. Boxbeam to column connections.

Figure 8-19. Cracks near shear studs.

8-10. Floor systems

a. *Floorbeams.* See figure 8-20.

(1) The end connections of floorbeams should be checked carefully for corrosion. This is particularly critical on truss bridges where the end connections are exposed to moisture and deicing chemicals from the roadway.

(2) The top flange of floorbeams should be checked for corrosion especially near the end connections and at points of bearing.

(3) The floorbeams should be checked to determine if they are twisted or swayed. This situation occasionally develops as a result of the longitudinal forces that are exerted by moving vehicles on

(b) Measure to reference point to determine section loss.

(c) Compare to plans or previous inspection notes.

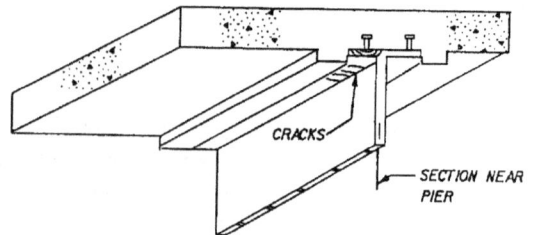

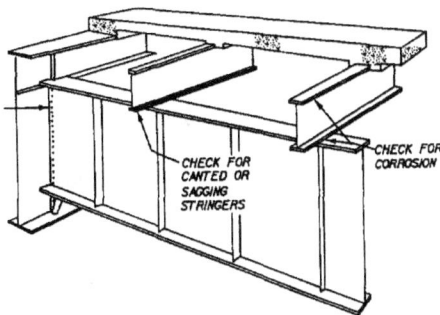

Figure 8-20. Steel floorbeam checklist items.

the floorbeams. It occurs primarily on older structures where the floorbeams are simply supported and where the stringers rest upon the floorbeams.

(4) The connections on the end floorbeams should be examined thoroughly for cracks in the welds and for slipped rivets or bolts.

b. *Stringers.*

(1) *Steel.*

(a) Check for rust or deterioration, especially around the top flange where moisture may accumulate from the floor above and at the end connections around rivets, bolts, and bearings.

(b) Check for sagging or canted stringers.

(c) Inspect all stringer connections for loose fasteners and clip angles (figure 8-21). Where stringers are seated on clip angles, check for cracks in the floorbeam web.

(2) *Timber.*

(a) Check for crushing and decay, especially along the top where the decking comes in contact with the stringer and at points at which the stringer bears directly upon the abutment and bent caps.

(b) Check for horizontal cracks and splitting, especially at the ends of stringers, where they are often notched.

(c) Check for sagging or canted stringers.

(d) Check the bridging between the stringers to determine whether it is tight and functioning properly.

(e) Check for accumulations of dirt and debris.

8-11. Diaphragms and cross frames

a. *Steel.*

(1) Check for loose or broken connections between the web of the beam or girder and the diaphragm (figure 8-22).

(2) Check for rust or other deterioration, especially around rivets and bolts, and those portions of the end of the diaphragms which come in contact with the bridge floor. These may be partic-

Figure 8-21. Clip angle stringer connection

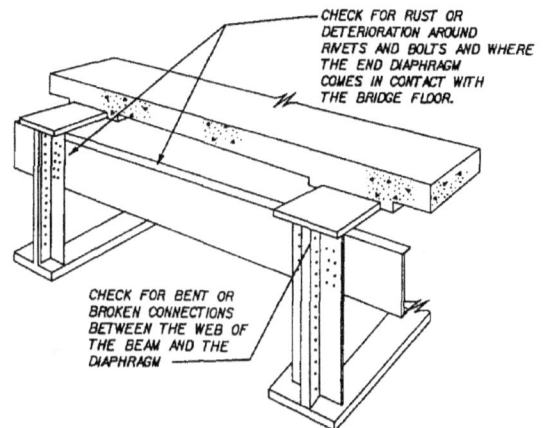

Figure 8-22. Diaphragm checklist items.

ularly susceptible to corrosion from roadway moisture and from deicing agents.

(3) Look for buckled or twisted cross frames. This situation may be an indication of overstress of the bracing.

b. *Timber.*

(1) Check for cracking or splitting, especially in end diaphragms that are supporting the floor.

(2) Check for decay along the top of the diaphragms where they may come in contact with the floor.

c. *Concrete.* Check for cracks, spalls, and for other forms of deterioration.

8-12. Trusses

a. *Steel trusses.*

(1) *Rust and deterioration.* On through trusses, moisture and deicing chemicals from the roadway are often splashed on the lower chord members and the member adjacent to the curb. The moisture and chemicals are retained at the

connection and between the adjacent faces of eye-bars, pin plates, etc. leading to rapid deterioration of the member. On riveted trusses, the horizontal surfaces and connections of lower chord members (figure 8-23) are especially susceptible to corrosion. Debris tends to accumulate causing moisture and salt to be retained. Note any deformation caused by expanding rust on the inside surfaces of laminated or overlapping plates.

(2) *Alignment of truss members.* End posts and interior members are vulnerable to collision damage from passing vehicles. Buckled, torn, or misaligned members may severely reduce the load-carrying capacity of the truss. Misalignment can be detected by sighting along the roadway rail or curb and along the truss chord members. Investigate and report any abnormal deviation.

(3) *Over-stressed members.* Local buckling indicates overstress of a compression member. Wrinkles or waves in the flanges, webs, or cover plates are common forms of buckling. Overstress of a ductile tension member could result in localized contraction in the cross section area of the member. This is usually accompanied by flaking of the paint.

(4) *Loose connections.* Cracks in the paint or displaced paint scabs around the joints and seams of gusset plates and other riveted or bolted connections may indicate looseness or slippage in the joints. Check rivets and bolts that appear defective.

(5) *Pins.* Inspect pins for scouring and other signs of wear. Be sure that spacers, nuts, retaining

caps, and keys are in place. Refer to paragraph 8-9 for a detailed inspection procedure.

(6) *Noise.* Note clashing of metal with the passage of live loads.

(7) *Riveted or bolted tension members.*

(a) Check each component to see that the loads are being evenly distributed between them by attempting to vibrate the members by hand and that batten plates are tight. If the loads are being unevenly distributed, one component might be loose or not have the right ring to it when struck with a hammer.

(b) Check carefully along the first row of bolts or rivets for cracking as the first row carries more load than succeeding rows. The first row is that closest to the edge of the gusset plate and perpendicular to the axis of the member.

(c) Check for nicks, gouges, and tears due to impact from passing vehicular traffic. This type of damage can initiate future cracks.

(d) Observe carefully any tack welding used either in construction or repair since this is a potential source of cracks. Any tack welds should be specially noted in the report for future observation and consideration in stress rating.

(e) If any misplaced holes or holes used for reconstruction have been plug welded, check carefully for fatigue cracks.

(8) *Welded tension members.*

(a) Check the full length of all longitudinal welds of each tension member for cracks.

(b) Check all joints at the ends of the members, including gussets.

(c) Check all transverse welding including internal diaphragms in box members.

(d) If connections are welded at gusset plates, carefully check these welds, particularly if any eccentricities observable by eye are involved.

(e) As with bolted or riveted members, check carefully for nicks, gouges, and tears due to impact damage and for any repairs made using tack welding.

(f) Box sections or other sections welded with backup bars should be checked carefully for discontinuity in the backup bars.

(g) Portions of fracture critical tension members which are difficult to access must be checked for corrosion using mirrors, fiberscopes, or boroscopes.

(h) Members should be examined carefully for any sites of arc strikes.

(i) Check carefully any holes that have been filled with weld metal since those are a source of fatigue cracking.

(9) *Eyebar members.* Whether or not these bars are fracture critical is dependent upon the

Figure 8-23. Lower chord of a riveted truss.

number of eyebars per member. During the inspection process, the inspector should:

(a) Inspect carefully the area around the eye and the shank for cracks (figure 8-24). This is where most failures occur in eyebars.

(b) Examine the spacers on the pins to be sure they are holding the eyebars in their proper position.

(c) Examine closely spaced eyebars at the pin for corrosion buildup (pack rust). These areas do not always receive proper maintenance due to their inaccessibility.

(d) Evaluate weld repairs closely.

(e) Check to determine if any eyebars are loose (unequal load distribution) or if they are frozen at the ends (no rotation).

(f) Check for any unauthorized welds and include their locations in the report so that the severity of their effect on the member may be assessed.

(10) Counters.

(a) Check the looped rod for cracks where the loop is formed.

(b) Observe the counters under live load for abnormal rubbing where the counters cross, and check this area carefully for wear (figure 8-25).

(c) Examine the threaded rods in the area of the turnbuckle for corrosion and wear.

(d) Test the tension in each rod to be sure they are not over-tightened or undertightened. The relative tension can be checked by pulling transversely by hand. The inspector should not adjust the turnbuckle but report the problem.

b. Timber trusses.

(1) Check for weathering, checking, splitting, and decay. Decay is often found at joints, caps, and around bolts holes. Decay is also common on the bridge seat.

(2) Check for crushing at the ends of compression chord and diagonal members.

(3) Examine splices carefully for decay. Note whether bolts and connections are tight.

(4) Check for decay at joints where there are contact surfaces, caps, where moisture can enter, and around holes through which truss bolts are fitted.

(5) Check end panel joints for decay.

(6) Check for dirt or debris accumulation on the bridge seat.

(7) Investigate the roof and sides of covered bridges for adequacy of protection afforded the structure members from the elements of weather.

(8) Check the alignment of the truss. Sagging of the truss may be due to the partial failure of joints or improper adjustment of steel vertical rods.

(9) Be particularly aware of fire hazards under the bridge, such as:

(a) Brush or drift accumulating.

(b) Storage of combustible material.

(c) Parking of vehicles.

(d) Signs of fires built.

Figure 8-24. Broken eyebar.

Figure 8-25. Worn counters due to rubbing.

8-13. lateral bracing portals and sway frames

a. Check all bracing members for rust, especially on horizontal surfaces such as those of lateral gusset plates and pockets without drains or with clogged drains.

b. Check for rust around bolts and rivet heads.

c. Look for loose or broken connections.

d. Check all upper and lower bracing members to observe whether they are properly adjusted and functioning satisfactorily.

e. Check for bent or twisted members. Since most of these bracing members work in compression, bends or kinks could significantly reduce their effectiveness. Since portals and sway braces necessarily restrict clearances, they are particularly vulnerable to damage from high loads.

f. Where lateral bracing is welded to girder flanges, inspect the weld and flanges for cracking.

g. Observe transverse vibration or movement of the structure under traffic to determine adequacy of lateral and sway bracing.

8-14. Tied arches

The stability of these types of arches is dependent upon the structural tie which is in tension. Therefore, the tie is always classified as a fracture critical member. The tie is the box member stretching horizontally between bearings. The majority of tied arch bridges built in the last decade have experienced some problems in the tie.

a. *Riveted or bolted members.* As with any FCM, the advantage of a riveted or bolted member is that they are internally redundant. Inspection should include:

(1) Observe built-up members to assure that the load is being evenly distributed to all components and that batten plates, lacing, and ties are tight. If members are loose or do not ring properly when struck with a hammer, loads may be distributed unevenly.

(2) Check the bolts or rivets at all connections (hangers, floorbeams, and end reactions) for cracks. All tack welds and/or strikes should be brought to the attention of the engineer for proper evaluation.

(3) Observe if any repairs or construction techniques have made use of tack welding and check carefully for cracks. All tack welds and/or strikes should be brought to the attention of the engineer for proper evaluation.

(4) Check the area around the floorbeam connection for cracks due to out-of-plane bending of the floorbeam and for cracks in rivet and bolt heads due to prying action.

(5) Check for corrosion sites with potential loss of section. This is particularly important in the inside of box-shaped members.

b. *Welded members.* These members must be thoroughly inspected from the inside as well as the outside. The key locations in the tie girder are the floorbeam connections, the hanger connections, and the knuckle area (area at the intersection of the tie girder and the arch rib). The inspection steps should include the following:

(1) Check all welds carefully for the entire length of the member. This applies primarily to the corner welds where the web and flange plates

are joined. Depending on the results of the corner weld inspection, it may be desirable to remove the backing bars and reexamine the welds. All fillet welds inside the girder should be inspected visually as accessible. Wire brushes should be used to clean the welds as necessary. The inspector should look for triaxial intersecting welds, irregular weld profiles, and possible intermittent fillet welds along the backup bars.

(2) Locate and inspect all the internal diaphragms and transverse butt welds. It may be necessary to clean the welds using a power wire brush.

(3) Check transverse connections at floorbeam with particular care. The usual location of the crack is near the corners, particularly if there is any gap between the floorbeam diaphragm and the web plates. This is a good place to clean with a power wire brush and use a dye penetrant test to ascertain the presence of cracks.

(4) If the box section has been welded with backup bars in the corners, as is often the case, the backup bars should be carefully examined for any breaks or poor splices.

(5) Portions of members that are difficult to access must be checked for corrosion using mirror, fiberscopes, or boroscopes.

(6) Hanger connections should each receive a thorough and detailed check. The purpose is to locate cracks or local distortions and to evaluate the extent of rusting or deterioration. These connections are where the support from the arch rib connects to the tie and, ordinarily, the floorbeams at the same location.

(7) The knuckle area at the intersection of the arch tie and arch rib is extremely complicated structurally and physically. Considerable study may be necessary to determine how to inspect all of the necessary locations in this area. Again, it may be necessary to use mirrors or devices. Also, dye penetrant testing should be used in this area if any suspicious crack-like formations are observed.

(8) At floorbeam connections and any splice points where the splices have been made with bolts, all of the bolts should be checked for tightness.

8-15. Metal bearings

In examining these types of bearings, determine initially whether they are actually performing the functions for which they have been designed. Bearings should be carefully examined after unusual occurrences such as heavy traffic damage, earthquakes, or batterings from debris in flood periods. Bearings should be inspected for the following (refer to figure 8-26):

a. Check to ensure that rockers, pins, and rollers are free of corrosion and debris. Excessive corrosion may cause the bearing to "freeze" or lock and become incapable of movement. When movement of expansion bearings is inhibited, temperature forces can reach enormous values.

b. Check rocker bearings where slots are provided for anchor bolts to ensure that the bolt is not frozen to the bearing.

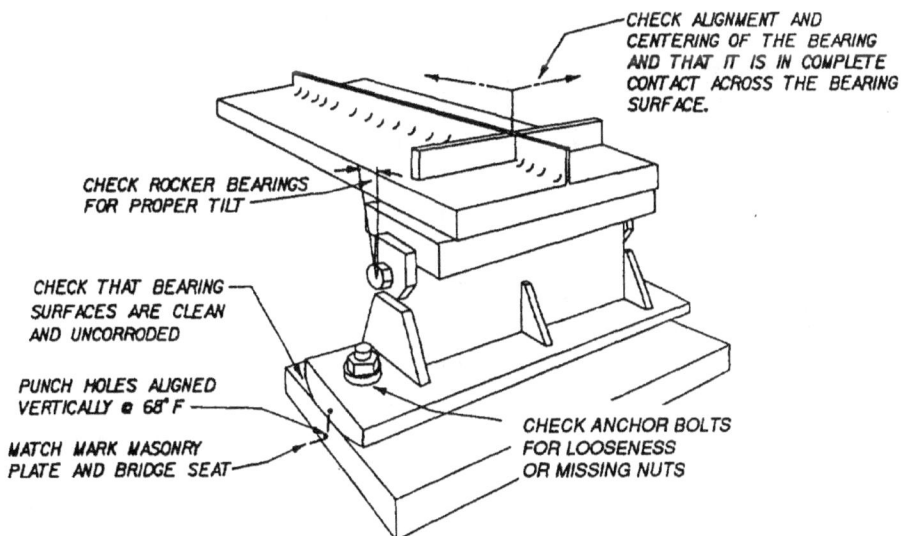

Figure 8-26. Metal bearing checklist items,

c. Check for dirt and corrosion on the bearing surfaces of rockers and rollers and the deflection slots around pins.

d. Determine whether the bearings are in proper alignment, in complete contact across the bearings surface, and that the bearing surfaces are clean.

e. Check barriers that require lubricants for proper functioning to ensure adequate and proper lubrication.

f. In those cases where bronze sliding plates are used, look for signs of electrolytic corrosion between the bronze and steel plates. This condition is common on bridges that are located in salt-air environments.

g. Detection of bearing rattles under live load conditions usually indicates that the bearings are loose. Determine the cause of this condition.

h. Check anchor bolts for looseness or missing nuts.

i. Measure the rocker tilt to the nearest 1/8-inch offset from the reference line as shown in figure 8-26. The appropriate amount of rocker tilt depends upon the temperature at the time of observation. Most rockers are set to be vertical at 68°F for steel bridges. Record the temperature at the time of inspection.

j. Measure the horizontal travel of the sliding bearings to the nearest 1/8 inch from the reference point. The two punch holes are aligned vertically at the standard of temperature used (usually 68°F). Record the temperature at the time of inspection.

k. On skewed bridges, bearings and lateral shear keys should be checked to determine if either are binding or if they have suffered damage from the creep effect of the bridge.

8-16. Elastomeric bearings

a. Check for splitting or tearing either vertically or horizontally. This is often due to inferior quality pads (figure 8-27).

b. Check for bulging caused by excessive compression. This may be the result of poor material composition.

c. Check for variable thickness other than that which is due to the normal rotation of the bearing.

d. Note the physical condition of the bearing pads and any abnormal flattening which may indicate overloading or excessive unevenness of loading.

8-17. Decks

a. Concrete decks. Check for cracking, scaling, and spalling of the concrete and record the extent of the deterioration. Refer to chapter 5 for guidance in recognizing and describing concrete deterioration.

(1) *Deck surface.* Note the type, size, and location of any deck deterioration.

(2) *Deck underside.* Inspect the underside of the deck for deterioration and water leakage. The passage of water through the deck usually causes some leaching of the concrete which forms grayish-white deposits of calcium hydroxide in the area of the leak known as efflorescence (figure 8-28). Extensive water leakage may indicate segregated or porous concrete or a general deterioration of the deck. Areas of wet concrete are additional indications of defective concrete.

(3) *Wearing surface.* Examine the wearing surface covering the concrete deck for reflection crack-

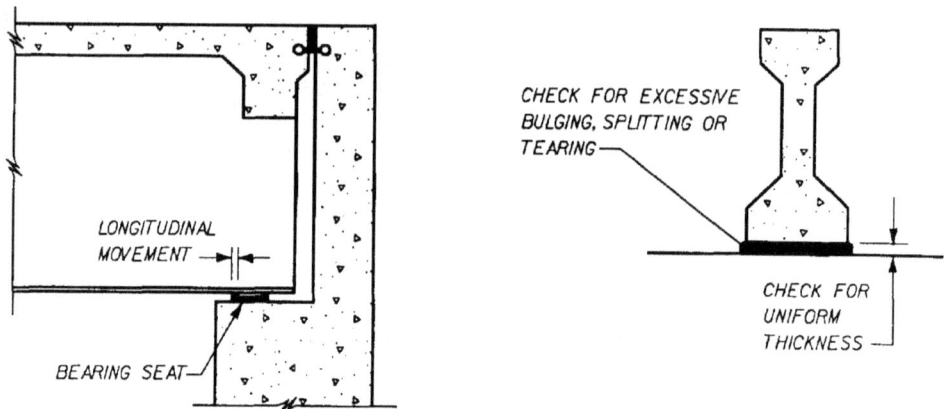

Figure 8-27. Elastomeric pad checklist items

ing and for poor adherence to the concrete. Deteriorated concrete beneath the wearing surface will often be reflected through the surface in the form of map cracking. Poor adherence leads to development of potholes. If deterioration is suspected, remove a small section of the wearing surface to check the condition of the concrete deck.

(4) *Wear.* Determine whether the concrete surface is worn or polished. When softer limestone aggregates are used in the concrete, fine aggregates and paste will be worn away, exposing the surface of the coarse aggregates to the polishing action of rubber tires. The resulting slippery surface becomes increasingly hazardous when the surface of the limestone is wet.

(5) *Stay-in-place forms.* If deterioration is suspected, remove several panels of the forms to permit examination of the underside of the deck. Rusty forms (figure 8-29), water dripping from pinholes in the form, or the separation of portions of the forms from the deck are reliable indications of deck cracking.

Figure 8-28. Efflorescence on the underside of a concrete deck.

Figure 8-29. Rusted stay-in-place forms underneath a concrete deck.

(6) *Reinforcing steel.* Note whether there are any stains on the concrete which would indicate that the reinforcing steel is rusting. Note whether any of the reinforcing steel is exposed.

b. *Timber decks.*

(1) *Deterioration.* Check for loose, broken, or worn planks, for loose fasteners, and for presence of decay particularly at the contact point with the stringer where moisture accumulates. Check asphalt overlays for the presence of potholes and cracking as a result of weak areas in the deck.

(2) *Traffic response.* Observe the timber deck under passing traffic for looseness or excessive deflection of the members.

(3) *Slipperiness.* Timber decks are sometimes slippery, especially when wet. Observe the traction of vehicles using the bridge for signs of this condition.

(4) *Utilities.* If utilities are supported by the bridge, note the effects on the bridge.

c. *Steel decks.*

(1) *General.* Check for corrosion and cracked welds. Maintenance of an impervious surface over a steel deck is an important safeguard against corrosion of the steel. Check to determine if the deck is securely fastened. Note any broken welds or clips. Determine if there is any loss of section due to rust or wear.

(2) *Slipperiness.* Note whether decks are slippery when wet.

(3) *Utilities.* If utilities are supported by the bridge, note any effect on the bridge.

d. *Open-grating decks.*

(1) *Cracks.* Examine the grating, support brackets, and stringers for cracking or welds.

(2) *Slipperiness.* Note whether decks are slippery when wet. Small steel studs may be welded to the grating to improve traction.

8-18. Expansion joints

a. Check all expansion joints for freedom of movement, proper clearance, and proper vertical alignment (figure 8-30). There should be sufficient room for expansion, but the joint should not be unduly open. Closed or widely opened joints or a bump at the back wall can result from substructure movements. Joints should be cleaned, filled and sealed to prevent seepage of water into the subgrade. This seepage causes subgrade failure and allows earth or debris to plug the joints and prevent closing in hot weather. The crowding of abutments against the bridge ends is common and can cause severe damage to the bridge. Proper opening size depends on the season, the type of joint seals, the temperature range, and the amount of slab expansion that must be accommodated by the joints. Normal temperature is usually assumed to be 65 to 70°F. Table 8-1 lists some general data for various types of expansion joints. The expansion length in table 8-1 is the portion of deck or structure expansion that must be accommodated by the joints. This distance may extend from the end of the bridge to the nearest fixed bearing, or it may be the sum of the distance on both sides of the joint. Multiplying the expansion length by the differential between the temperature at the particular moment and 68°F and this product by 0.0000065 will give the approximate change in joint opening from the values listed. Very often, construction plans will give useful data concerning the setting of expansion devices.

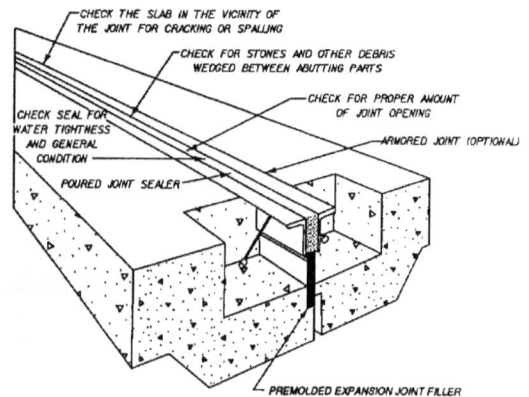

Figure 8-30. Expansion joint checklist items.

Table 8-1. Expansion joint data

Joints	Expansion Lengths	Joint Openings at 69°F
Steel finger dams	200-foot minimum	3 inches min.
Steel expansion plates	200-foot maximum	2 inches
Compression seals	135-foot maximum	1 5/8 inches
Poured sealants and joint fillers	120-foot maximum	1½ inches

b. Check seals for water tightness and general conditions. Look for:

(1) Seal or sealant pulling away from the edges of the joints.

(2) Abrasion, shriveling, or other physical deterioration of the seal.

(3) Stains and other signs of leakage underneath the deck. Leaking seals permit water and brine to flow onto the bridge seat and pier cap causing corrosion of the bearings, disintegration of the concrete, and staining. Joints not properly sealed should be cleaned and resealed.

c. Check to see that expansion joints are free of stones and other debris. Stones lodged in the joints can create localized stresses which may cause cracking and spalling of the deck. Large amounts of debris cause jamming, thus rendering the joints ineffective.

d. Examine steel finger-type joints and sliding plate joints for evidence of loose anchorages, cracking or breaking of welds, or other defective details. Sometimes the fingers may be damaged by traffic or by cracks which have developed at the base of the fingers.

e. Verify that surfacing material has not jammed the finger joints on bridges that have been resurfaced several times.

f. Examine specifically the underside of the expansion joint, regardless of accessibility, to detect any existing or potential problem.

g. Sound the concrete deck adjacent to all expansion devices for voids or laminations in the deck.

8-19. Railings, sidewalks, and curbs

a. Railings.

(1) Inspect all railings for damage caused by collision and for weakening caused by some form of deterioration.

(2) Check concrete railings for cracking, disintegration, and corrosion of rebars.

(3) Check steel and aluminum railings for loose posts or rails and for rust and other deterioration. In particular, check the condition of the connections of the posts to the decks, including the condition of the anchor bolts and the deck area around them.

(4) Check timber railings for decay, loose connections, and for missing or damaged rails.

(5) Check the vertical and horizontal alignment of all handrails for any indications of settlement in the substructure or any bearing deficiencies.

(6) Examine all handrail joints to see that they are open and functioning properly.

(7) Examine all handrails to see that they are of adequate height, secure, and relatively free of slivers or any projections which would be hazardous to pedestrians.

(8) Check for rust stains on the concrete around the perimeter of steel rail posts which are set in pockets. Remove grout from around the posts and determine severity of corrosion if rust stains indicate such action is warranted.

(9) Note whether barrier railings on the approaches to the bridge extend beyond the end of the bridge railing or parapet end and are anchored to the inside face (figure 8-31). This feature reduces the severity of vehicle collision. In situations where parapet ends are unprotected and no approach rail exists, a flared, tapered approach railing should be installed. On two-way bridges, this type of railing should be installed at both ends of the existing railings or parapet.

(10) Examine barrier railings for traffic damage and alignment.

(11) Check concrete barrier railings for cracks, spalls, and other deterioration.

(12) Check for corrosion in the metal portions of barrier railings and determine whether the anchor bolts and nuts are tight.

Figure 8-31. Unprotected parapet end of a bridge.

b. Sidewalks.

(1) Check concrete sidewalks and parapets in the same manner as the bridge decks for cracks, spalls, and other deteriorations.

(2) Examine the condition of concrete sidewalks at joints, especially at the abutments, for signs of differential movement which could open the joint.

(3) Check steel sidewalks for corrosion and to see that all connections are secure.

(4) Check timber sidewalks for soundness of the timber. Determine whether the floor planks are adequately supported.

(5) Check timber sidewalks for hazards to pedestrians such as loose or missing planks, large cracks, decay or warping of the planks, protruding nails, or other hazardous conditions.

(6) Check slickness of surfaced timber during wet or frosty weather conditions to determine whether any corrosive action is necessary.

(7) Check sidewalk drainage for adequate carryoff. Examine the sidewalk surface for roughness or other conditions that may make walking hazardous or difficult.

(8) Check structural integrity of sidewalk brackets.

c. Curbs.

(1) Check concrete curbs for cracks, spalls, and other deterioration.

(2) Check timber curbs for splitting, warping, and decay.

(3) Report any curbs or safety walks which project into the roadway or a narrow shoulder of the roadway, since they are safety hazards.

(4) Note any loss of curb height resulting from the buildup of the deck surface.

(5) Examine timber wheel guards including scupper blocks for splits, checks, and decay.

(6) Check timber wheel guards to see whether they are bolted securely in place.

(7) Note condition of the painting of timber wheel guards where paint is used to improve visibility.

8-20. Approaches

a. At the joint of the bridge backwall.

(1) *Vertical displacement.* Laying a straight edge across the joint will record any differences in elevation across the joint not caused by the grade. If the deck is lower than the approach, or if the straight edge indicates a rotation, then foundation settlement or movement may have occurred, and other indications of such action should be checked.

(2) *Joint width (horizontal displacement).*

(a) *Incorrect opening.* Measure the joint width for increased or decreased openings. Either condition indicates foundation movement. A decreased opening could also be caused by pavement thrust. Other parts of the bridge affected by such occurrences should also be investigated.

(b) *Clogged joints.* Where joints are clogged or jammed with stones and hard debris, the expansion joints will be unable to function properly, and pavement thrust will develop. Make particular note of this condition.

(c) *Joint seal.* The integrity of the joint sealant is critical to protecting the soil or portions of the bridge under the joint, particularly the bridge seat, from water. The seal may be damaged by either weathering, traffic abrasion, or movement of the seal itself.

b. Other transverse joints near the bridge. Examine these joints for closing or clogging, since they are liable to the same difficulties as the backwall joints. The inspector should note the relative movements (if any) of the joints, any clogging with stones or other debris, and any failure, deterioration, or slippage of the joint seal. The extent of these defects should be reported.

c. Approach slabs. Check for cracking or tipping of the approach slabs. These are indications of poor backfill compaction (although on a skewed bridge, it would not be unusual for an acute corner to crack).

d. Shoulders and drainage. Check the shoulders and determine whether they are maintained at the same height as the pavement. There should be adequate provisions to carry off drainage in the catch basins or ditches, especially if water is allowed to flow off the bridge deck (see figure 8-25).

e. Approach slopes. Check the approach slopes for adequacy and report any condition or other surface defects that make the approach unusually rough or indicate approach settlement.

f. Pavement approaches. Report any potholes, severe cracks, surface unevenness, or other surface defects that make the approach unusually rough or indicate approach settlement.

8-21. Bridge drainage

Almost all of the drainage problems encountered by an inspector are caused by the failure of the drainage system to carry water away. Poor deck drainage usually leads to deck disintegration. The following items should be checked:

a. Clogging or inadequate drainage openings.
Check the deck and the deck inlets for signs of
clogging or inadequate drainage openings. These
deficiencies will be manifested by debris on or
around the inlet after a storm. Scaling and concrete deterioration around the inlet are additional
signs of an inadequate inlet.

b. Water stains. Observe the bridge beams,
piers, and abutments for water stains. These may
indicate leaky pipes, filled gutters, or scupper
discharge pipes that are too short. However,
stained abutments and piers could also mean
leaky joints.

c. Drain outlets. Check to see that deckdrain
outlets (scuppers) do not discharge water where it
may be detrimental to other members of the
structure, cause fill and bank erosion, or spill onto
a roadway below.

d. Damaged pipes. Look for pipes damaged by
freezing, corrosion, or collision. These will show
cracks, holes, or stains.

e. Clogged pipes. Check pipes, if possible. Open
the cleanout at bottom of pipes to see whether
pipes are open all the way through.

f. Sand or soil accumulations. Check for layers
of sand or soil on the bridge deck. Presence of
either of these will retain moisture and brine and
will accelerate deck scaling. Soil or sand deposits
are clear indications of poor deck drainage (figure
8-32).

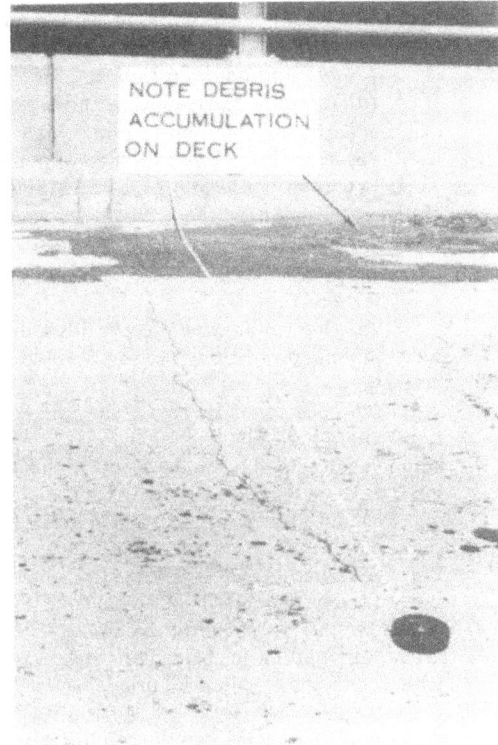

Figure 8-32. Debris accumulation on bridge deck indicating drainage problem.

Section III. MISCELLANEOUS INSPECTION ITEMS

8-22. Waterways

a. Maximum water level. Ideally, waterways
should be inspected during and immediately following periods of flood, since the effects of high
water will be most apparent at these times. Since
this is not always possible, a knowledge of the
heights of past major floods from stream gaug-
ing records or from other sources, together with
observations made during or immediately following high water, are helpful in determining the
adequacy of the waterway opening. Other sources
are:

(1) High water marks or ice scars left on
trees.

(2) Water marks on painted surfaces.

(3) Debris wedged beneath the deck of the
bridge or on the bridge seats.

(4) Information from established local resi-
dents.

b. Insufficient freeboard. This is a prime characteristic of inadequate waterways. In addition to
the signs mentioned previously, lateral displace-

ment of old superstructures is a prime indication
of insufficient freeboard.

c. Debris. Debris compounds the problems of a
scanty freeboard. Check for debris deposits along
the banks upstream and around the bridge.

d. Obstruction. Debris or vegetation in the waterways, both upstream and downstream, may
reduce the width of the waterway, contribute to
scour, and even become a fire hazard. Sand and
gravel bars formed in the channel may increase
stream velocity and lead to scour near piers and
abutments.

e. Scour.

(1) *Channel profile.* In streambeds susceptible
to scour and degradation, a channel profile should
be taken periodically. Generally, 100-foot intervals, extending to a few hundred feet upstream
and downstream, should be sufficient. This information, when compared with past records, will
often reveal such problems as scour, shifts in the
channel, and degradation.

(2) *Soundings.* Soundings for scour should be

taken in a radial pattern around the large river piers.

(3) *Shore and bank protection.*

(a) Examine the condition and adequacy of existing bank and shore protection.

(b) Check for bank or levee for erosion caused by improper location or skew of the bridge piers or abutments.

(c) Note whether channel changes are impairing or decreasing the effectiveness of the present protection.

(d) Determine whether it is advisable to add more channel protection or to revise the existing protection.

f. High backwater. Be particularly alert for locations where high fills and inadequate or debris-jammed culverts may create a very high backwater. The fill acts as a dam, and with the possibility of a washout during rainfall, a disastrous failure could result.

g. Wave *action.* Observe the effect of wave action on the bridge and its approaches.

h. Existing or potential problems. Observe the areas surrounding the bridge and its approaches for any existing or potential problems, such as ice jams.

i. Spur dikes. Observe the condition and functioning of existing spur dikes.

8-23. Paint

a. Examine all paint carefully for cracking or chipping, scaling, rust pimples, and chalking. Look for evidence of "alligatoring." If the paint film has disintegrated, note whether the prime coat or the surface of the metal is exposed. Note the extent and severity of the paint deterioration. If extensive "spot" painting will be required, probably the entire structure should be repainted; otherwise, spot painting will most likely be sufficient.

b. Look for paint failure on upper chord horizontal surfaces, or those surfaces which are most exposed to sunlight or moisture. Give particular attention to areas around rivets and bolts, the ends of beams, the seams of built-up members, the unwelded ends of stiffeners, and any other areas that are difficult to paint or that may retain moisture.

8-24. Signing

This section is concerned with the presence and effectiveness of bridge signing. Since some bridges on military installations must carry both military and civilian traffic, signing may be necessary for both types of traffic. The required regulatory signing for the bridge to be inspected should be determined prior to the inspection. The absence of required signing and the condition of the existing signs should be noted during the inspection. For civilian traffic, the AASHTO "Manual on Uniform Traffic Control Devices" should be consulted for specific information with regard to signing. Military signing should be inspected according to the following guidelines:

a. Type of signs. When inspecting a bridge for signing, not only should the presently posted signs be inspected, but it should also be determined whether additional signs are needed because of changed bridge or roadway conditions. The types of warning and regulatory signs normally required are:

(1) *Weight limit.* This is the most important inspection item, particularly for the older and deteriorated bridges. The weight limit should be determined by the bridge classification procedure outlined in chapter 9. Depending upon the bridge type and expected traffic, the bridge may need both military and civilian load classifications. Typical military load classification signs are shown in figure 8-33. Civilian load limit signs should be in accordance with the local legal requirements and with AASHTO "Manual on Uniform Traffic Control Devices."

(2) *Overhead clearance.* Minimum overhead clearances for military vehicles (as summarized in table 8-2) should be checked. When the overhead

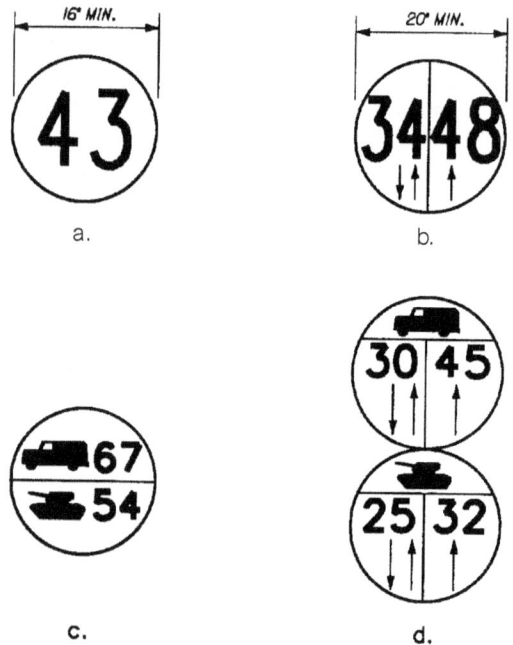

Figure 8-33. Typical military load-class signs.

Table 8-2. Minimum overhead clearances for bridges

Bridge Class	Minimum Overhead Clearance
4-70	14' 0"
71-over	15' 6"

clearance is less than the values prescribed in the table, the actual clearance must be indicated by the use of a telltale sign as shown in figure 8-34. Civilian traffic bridges with clearance restrictions should be marked as shown in figure 8-35. For civilian traffic, any clearance that is less than 1 foot higher than the local legal height and road limit should be posted with a "Low Clearance" sign. Where existing civilian signs are in place and are sufficiently clear, additional clearance signs (vertical or horizontal) for military vehicles are not required.

(3) *Roadway width.* Bridge width limitations often limit the maximum class of military vehicle for the bridge. Minimum lane widths for specific bridge classes are summarized in table 8-3 and should be checked during the inspection. Many new bridges are narrower than they should be. For civilian bridges, "Narrow Bridge" signs and striped paddleboards (figure 8-35) should be used when the bridge width is less than that of the approach roadway. If the superstructure or parapet end extends above the curb, it should be striped and a reflectorized hazard marker should be attached.

(4) *Narrow underpass.* Where the roadway narrows at an underpass or where there is a pier in the middle of the roadway, striped hazard

markings should be placed on the abutment walls and on pier edges of civilian bridges. Reflective hazard markers should also be placed on the piers and abutments, and the approaching pavement should be appropriately marked to warn the approaching traffic of the hazard.

(5) *Speed and traffic markers.* These types of signs should be checked to ascertain whether they are appropriate. Speed restrictions should be carefully noted to determine whether such restrictions are consistent with bridge and traffic conditions. Additional traffic markers may be required to facilitate the safe and continuous flow of traffic.

b. Details of military signs.

(1) *Circular signs.* Both military and civil bridges should be marked with circular signs indicating the military load classification. They should have a yellow background with black inscriptions, as large as the diameter of the sign allows. These signs should be placed at both ends of the bridge in a position that is clearly visible to all oncoming traffic. The required sizes and appearance for these signs are summarized in figure 8-33.

(2) *Rectangular signs.* Additional instructions and technical information are inscribed on rectangular signs. These signs should be a minimum of 16 inches in height or width and have a yellow background upon which the appropriate letters, figures, or symbols are inscribed in black. These signs, other than those indicating height restrictions, should be placed immediately below the bridge classification signs. Height restriction signs should be placed centrally on the overhead obstruction.

c. Details of civilian signs.

(1) *Location.* The weight limit sign, being regulatory, should be located just ahead of the

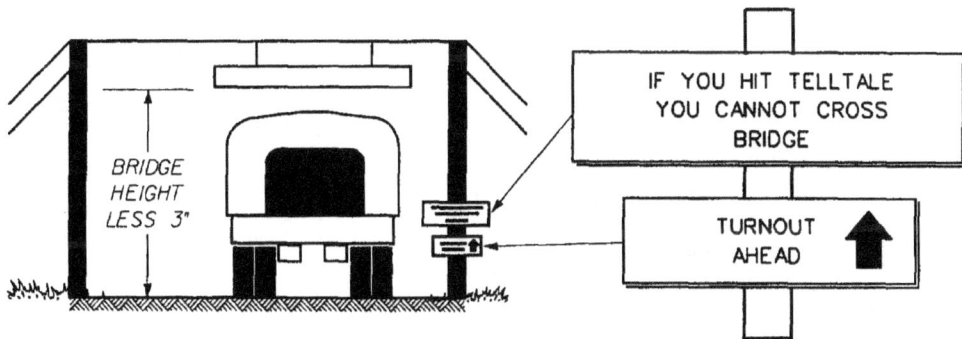

Figure 8-34. Typical telltale indicating overhead clearance of bridge.

Figure 8-35. Appropriate markings for clearances on civilian bridges.

Table 8-3. Minimum lane widths for bridges

Bridge Class	Minimum Width Between Curbs	
	One Lane	Two Lane
4-12	9' 0"	18' 0"
13-30	11' 0"	18' 0"
31-60	13' 2"	24' 0"
61-100	14' 9"	27' 0"

bridge. Lateral clearance of the sign should be determined by the requirements of the highway type. On heavily traveled roads (such as freeways), side-mounted signs should be:

(a) Positioned 30 feet from the edge of the travel way.

(b) Located behind a barrier, guardrail.

(c) Affixed to a breakaway installation. On less traveled roads, signs should preferably be located behind a barrier guardrail or be affixed to a breakaway standard. Any sign support of sufficient mass to be a hazard, which does not meet the preceding criteria, should be noted and reported.

(2) Condition. It is important that all caution signs be in good condition. Evaluation of the condition and adequacy of the signing will depend upon the conditions prevailing in a given area. It is suggested that the AASHTO "Manual on Uniform Traffic Control Devices" be consulted for specific information with regard to signing. Signs should be checked for:

(a) Reflectorization. Adequate reflectoriza-tion and/or painting are required for night visibil-ity.

(b) Legibility. Note whether the legend is difficult to read. This may be because of dirt encrustment, dulled paint, inadequate lettering, or inadequate sign size. Refer to the AASHTO "Manual on Uniform Traffic Control Devices" for guidelines as to criteria to be followed in evaluat-ing sign legibility.

(c) Vandalism. Bullet holes, paint smears, campaign stickers, etc. should be noted.

(d) Minimum sizes. In general, most warn-ing signs are diamond-shaped and measure at least 30 by 30 inches. "Low Clearance" signs are 36 by 36 inches. The "Weight Limit" signs are rectangular with minimum dimensions of 18 by 24 inches.

(e) Vegetation. On minor roads, heavy vege-tation growth may obscure signs. Note the type and location of such vegetation so that it may be trimmed or removed entirely. If relocation of the sign(s) is necessary, include such remarks in the inspection report.

(f) Sign support damage. Note whether sign supports are bent, twisted, or otherwise damaged.

8-25. Utilities

a. Check pipe, ducts, etc., for leaks, breaks, cracks, and deteriorating coverings.

b. Check the supports for signs of corrosion, damage, loose connections, and general lack of rigidity. If utility mounts rattle during passage of traffic, especially on steel bridges, note need for padding.

c. Check the annual space between pipe and sleeve or between the pipe and the blocked-up area for leaks where utilities pass through abutments.

d. Check for leaky water or sewer pipes for damage.

e. Inspect the area under water or sewer pipes for damage.

f. Determine whether mutually hazardous transmittants, such as volatile fuels and electricity, are sufficiently isolated from each other. If such utilities are side-by-side or in the same bay, report this condition for auxiliary encasement or future relocation.

g. Check utilities that are located beneath the bridge for adequate roadway clearances.

h. Determine whether any utility obstructs the waterway area or is positioned so that it hinders drift removal during periods of high water.

i. Check the encasement of pipes carrying fluids under pressure for damage, and check vents or drains for leaks.

j. Check for the presence of shutoff valves on pipelines carrying hazardous pressurized fluids, unless the fluid supply is controlled by automatic devices.

k. Note whether any utility is located where there is a possibility that it may be struck and damaged by traffic or by ice and debris carried by high water.

l. Determine whether utilities are adequately supported and whether they present a hazard to any traffic which may use or pass under the bridge.

m. Check for wear or deteriorated shielding and insulation on power cables.

n. Check for adverse effect utilities may have on the bridge, e.g., interferences with bridge maintenance operations or an impairing of structural integrity.

o. Check whether vibration or expansion movements are causing cracking in the support members.

p. Check supporting members of the bridge for paint damage.

q. Note any adverse aesthetic effect utilities may have on the bridge.

8-26. Lighting

a. Whitewuy lighting.

(1) *Collision.* Note any light poles that are dented, scraped, cracked, inclined, or otherwise damaged.

(2) *Fatigue.* Aluminum light standards and castings are most likely to suffer from fatigue. Check for cracking in:

(a) Mast arms and cast fitting on standards.

(b) At the base of standards, especially the cast elements.

(3) *Corrosion.* Check steel standards for rusting and concrete standards for cracking and spalling.

b. Electrical systems. This part of the inspection should be made by or with the assistance of a qualified electrician.

(1) *Wiring.* Observe any exposed wiring for signs of faulty, worn, or damaged insulations. Note and report the following:

(a) Bad wiring practices.

(b) Bunches of excess wires.

(c) Loose wires.

(d) Poor wire splices.

(e) Inadequate securing of groundlines.

(2) *Junction boxes.* Check inside junction boxes for excessive moisture, drain hole, poor wire splices, and loose connections. Note the condition of wiring and insulation. Where the base of the light standards contains a junction box, examine this as well. Note whether the junction box, outlet box, or switch box covers are in place.

(3) *Conduit.* Check conduits for rust or missing sections. Check the curbs and sidewalks for large cracks that might have fractured the conduit imbedded in them. Note whether the conduit braces and boxes are properly secured.

(4) *Whiteway current.* On those structures where the whiteway current is carried over an open line above the sidewalks, check for hanging objects such as fishing lines and moss.

c. Lamps or damaged standards. Note any missing or damaged standards. A cover placed over the electric eye controller will turn on the lights. Note the number and the locations of those lights that do not illuminate.

d. Sign lighting. Inspect sign lighting for the same defect as conventional lighting.

e. Navigation lights.

(1) *Lights.* Check to determine whether all of the required navigation lights are present and properly located. For fixed bridges, a green navigation light is suspended from the superstructure over the channel centerline and red lights are placed to make the channel edges. When piers are situated at the channel edges, the red lights are positioned on the piers or fenders. For movable spans, navigation lighting requirements vary according to the bridge type. When in doubt as to the requirements, refer to Section 68 of the Coast Guard Pamphlet CG204, "Aids to Navigation."

(2) *Lighting devices.* Check the overall condition of lighting and devices to determine whether they are rusted, whether any of the lenses are

broken or missing, and whether the lights are functioning correctly.

(3) *Wiring*. Check the condition of wiring, conduits, and securing devices to determine whether they are loose or corroded.

f. Aerial obstruction lights. For short bridges (less than 150 feet), there should be at least two continuous glow red lights mounted at the high points of the superstructure. For longer structures, the red lights should be placed at 150- to 600-foot intervals, while sets of at least three flashing red beacons should be mounted atop the peaks of widely separated high points such as suspension bridge towers, truss cusps, etc. Check these lights for proper maintenance and functioning.

Section IV. INSPECTION OF RAILROAD BRIDGES

8-27. General

a. The construction of railroad bridges is the same as that for all of the previously discussed roadway bridges. Their construction may also be from steel, concrete, masonry, or wood. As a result, most aspects of their inspection are the same, including inspector qualifications, frequency of inspection, and the required thoroughness of inspection.

b. Thorough inspection of the track portion of railroad bridges is usually conducted separately and by different personnel and inspection standards. Therefore, track inspection is only briefly covered in this manual. The primary emphasis for the bridge inspector should be on the supporting bridge structure itself.

c. If, during the inspection, the inspector finds any condition that he considers serious enough to possibly result in collapse of the bridge, he should immediately close the bridge and notify the proper authorities.

8-28. Railroad deck types

Railroad bridge decks are generally of two different types: open deck and ballast deck. An open deck bridge has bolts securing the ties to stringers which in turn are attached to a pile bent or pier cap. On a ballast deck bridge, the deck is a solid floor with a regular ballast section placed on top of the floor.

8-29. Track inspection

As previously noted, a thorough track inspection is generally performed separately from the bridge inspection. However, a thorough bridge inspection should include a minimal inspection of the track since track defects can adversely affect the integrity of the bridge structure itself. The following items relating to the track should be inspected as a minimum:

a. Check the alignment of the track, both vertically and horizontally. State whether the track is level or on grade and if alignment is tangent or curved. If on a curve, note how the superelevation is provided, whether by cutoff in the bents or by taper in the caps or in the ballast section. Note the location of the track with reference to the chords for uniformity of loading.

b. Where track appears out of line or surface, note the location, degree of misalignment, and the probable cause.

c. Check the condition of the embankment at the bridge ends for fullness of crown, steepness of slopes, and depth of bulkheads. Note whether the track ties are fully ballasted and well bedded.

d. Record the weight and condition of the track rails and inside guardrails. Check the condition of any rail joints, fastenings, and tie plates.

e. Any rail anchors found on track over open deck bridges are to be removed immediately. Anchors on open deck bridges are particularly dangerous because the ties form an integral part of the bridge structure. Should the rail start to run, the rail anchors will put longitudinal forces into the ties which will be transferred to the remainder of the bridge structure, possibly causing structural damage. Where anchors are used on track approaching open deck bridges, every third tie should be box anchored (four anchors per tie) for at least two rail lengths off each end of the bridge. Very importantly, no anchors are to be applied on the bridge itself.

8-30. Deck inspection

a. Guardrails are installed to guide derailed equipment and prevent it from leaving the track. All bridges and their approaches should be equipped with guardrails which extend at least 50 feet past the ends of the structure. The condition of the guardrails should also be closely inspected. Look for loose or missing spikes, joint bars, track bolts, or tie plates.

b. On ballast deck bridges, check the ballast to see if it is clean and in full section. The ballast should be measured from the base of the rail at each end of bridge. The ballast section should be clean, free-draining, and free of vegetation, soil (mud), and other foreign materials. The ballast materials should not be at a level above the top of the ties.

c. Ballast deck bridges are normally equipped with drainholes to allow water to drain from the bridge deck. Check these holes to make sure they are open and free draining.

d. Walkways or walkboards along an open deck bridge should be maintained to allow for safe walking over the structure. Check for loose, broken, deteriorated, or missing boards.

e. The ties should be inspected as follows:

(1) Note the size, spacing, and uniformity of bearing of all ties. In a ballast deck bridge, make sure all of the ties are fully ballasted and well bedded. Bolts that secure the ties to the bridge stringers in open deck bridges should be checked for deterioration and sufficient tightness. Any tie that is not materially defective (paragraph 8-30e(2)), but does not fully support both rails, should be noted and recommended for tamping up and respiking.

(2) Typical tie defects are shown in figure 8-36 and a tie should be considered defective (and noted) if it is:

(a) Broken through.

(b) Split or otherwise impaired to the extent that it will not hold spikes or other rail fasteners.

(c) So deteriorated that the tie plate can move laterally more than ½ inch relative to the crosstie.

(d) Cut by the tie plate more than 2 inches.

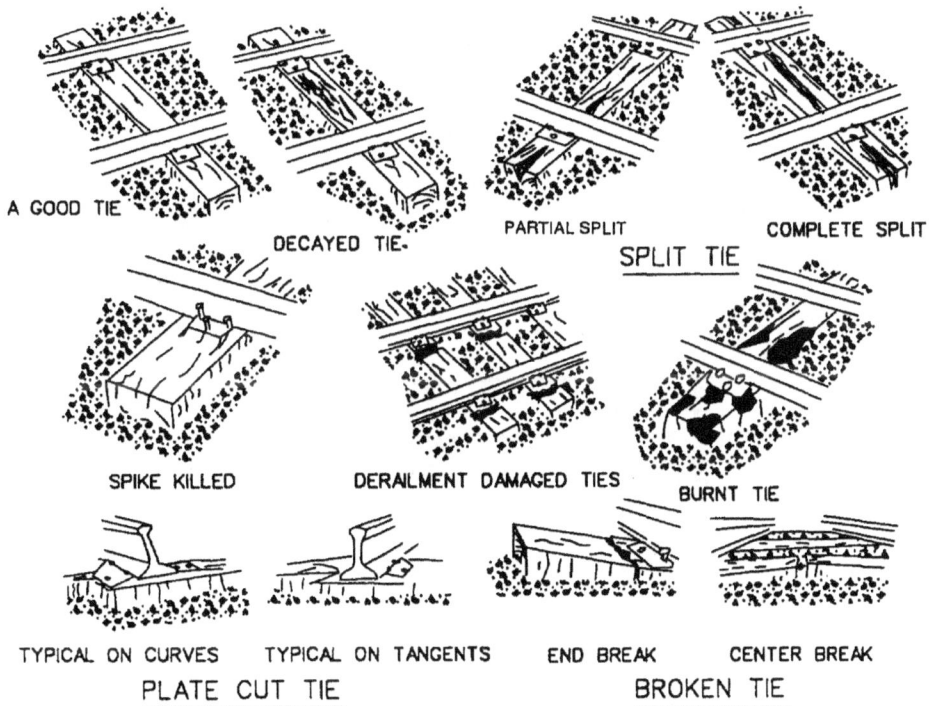

Figure 6-36. Examples of good and defective cross-ties.

(e) Cut by wheel flanges, dragging equipment, fire, etc. to a depth of more than 2 inches within 12 inches of the base of the rail, frog, or load-bearing area.

(f) Rotted, hollow, or generally deteriorated to a point where a substantial amount of the material is decayed or missing.

(3) The occurrence of consecutive defective ties in categories 1 and/or 2 requires operating restrictions as specified in table 8-4.

(4) All track joints should be supported by at least one nondefective tie whose centerline is within 18 inches of the rail ends as shown in figure 8-37. At any location where a rail joint is not supported by at least one nondefective tie, operations should not exceed 10 miles per hour (mph).

(5) If the existing tie spacing averages greater than 22 inches within the distance of a rail length, the desired spacing should be established during the next major maintenance cycle. For track constructed with an average tie spacing greater than 22 inches, the desired spacing should be established during the next track rehabilitation.

(6) Missing or skewed (crooked) ties are undesirable in track. At any location where the center-to-center tie spacing measured along either rail exceeds 48 inches, operations shall not exceed 10 mph until additional tie support is provided, or skewed ties should be straightened during the next track rehabilitation.

8-31. Superstructure inspection

The inspection procedures for the superstructure portion of railroad bridges are the same as those

Table 8-4. Operating restrictions

Number of Consecutive Defective Ties	Operating Restrictions
0 to 2	None
3	Limit maximum speed to 10 mph
4	Limit maximum speed to 5 mph
5 or more	No operation

previously discussed for roadway bridges with the following exceptions:

a. When possible, the movement of the superstructure during passage of a train should be observed. Note excessive movements, rattles, and vibrations.

b. Observe all members to determine if any are broken or moved out of proper position and whether all fastening devices are functioning properly.

c. Check all stringers for soundness and surface defects. Note their size and type and the number used in each panel. Note if the bearing is sound and uniform, if all stringers are properly chorded and securely anchored, and if all shims and blocking are properly installed. Note whether packers or separators are used and the condition of all chord bolts.

d. With timber trestles, fire protection is very important. The following items should be inspected:

(1) Note whether the surface of the ground around and beneath the structure is kept clean of grass, weeds, drift, or other combustible material.

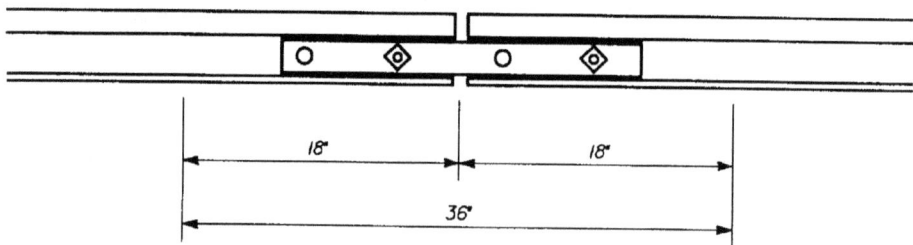

AT EACH JOINT. AT LEAST ONE TIE WITHIN THIS AREA MUST BE NON-DEFECTIVE.

Figure 8-37. Required tie support at track joints.

(2) Where rust-resisting sheet metal is used as a fire protection covering for deck members, note the condition of the metal and its fastenings.

(3) Note if any other method of fire protection has been used, such as fire retardant salts, external or surface protective coatings, or fire walls. Record such apparent observations as are pertinent to the physical condition and effectiveness of such protective applications.

(4) Where water barrels are provided, note the number, condition, if filled, and if buckets for bailing are on hand. If sand is used, note whether bins are full and in condition to keep the sand dry.

(5) Note if timber, particularly top surfaces of ties and stringers in open deck bridges, is free from frayed fiber, punk wood, or numerous checks.

8-32. Substructure inspection

The inspection procedures for the substructure portion of railroad bridges are the same as those previously discussed for roadway bridges with the following exceptions:

a. When possible, the movement of the substructure during passage of a train should be observed. Note excessive vibration, deflection, side sway, and movement at pier supports.

b. Examine all bents and towers for plumbness, settlement, sliding and churning, and give an accurate description of the nature and extent of any irregularities. Note particularly whether caps and sill have full and uniform bearing on the supports.

c. Note the number and kind of piles or posts in the bents or towers. Note the uniformity of spacing and the location of any stubbed or spliced members, especially if the bridge is on a curve or the bent is more than 15 feet in height.

d. Check all fastening devices for physical condition and tightness.

8-33. Recommended practices

The inspector's outline of repairs should be based on the following recommended practices. Refer to chapters 11 through 13 of this manual for specific details for these repairs:

a. Posting of the outside piles should not be permitted on bridges on curves where bents exceed 12 feet in height or on tangents where bents are over 20 feet in height.

b. On high-speed track where traffic is heavy, not more than two posted piles in any one bent shall be permitted. If more than two piles are poor, all piles should be cut off to sound wood below groundline and a framed bent installed or piles redriven.

c. All posts should be boxed, in addition to toe nailing, to prevent buckling.

d. When individual caps, sills, braces, or struts have become weakened beyond their ability to perform their intended function, renewal is the only remedy.

e. When only an individual stringer is materially deteriorated, an additional stringer may be installed, inside or outside of the chord, to aid the weakened member.

f. Where piles are decayed at the top they may be cut off and double capped; a single pile may be corbelled.

g. Shimming of stringer to provide proper surface and cross level should be done with a single shim under each chord. If possible, avoid multiple shimming.

Section V. BOX CULVERTS

8-34. Types of distress

A culvert is generally used where its construction would permit a fill to substitute for a bridge without any loss of vital waterway area. This combination of high earth loads, long pipe-like structures, and running water tends to produce the following types of distress:

a. The basic causes and actions of foundation movements are discussed in chapter 5. Here, they need only be listed:

(1) Settlement of the box. This may be either a smooth sag, or it may be differential settlement at the expansion joints.

(2) Tipping of wing walls.

(3) Lateral movements of sections of the box.

b. High embankments may impose very heavy loads on the top and bottom slabs. These earth pressures can cause either shear or flexural failures in the top slabs.

c. Construction defects can lead to structural distress.

d. Undermining is a form of scour attack on the upstream and downstream ends of box culverts. When sheeting or a concrete cut-off wall is not provided or is not deep enough, the stream may wash away the soil under the ends of the floor slab, the apron, or the wing wall footings, leading to settlements and culvert cracking.

e. Plugging may result from debris collecting over the mouth of the culvert. This can cause flooding and flotation and displacement of part, or all, of the box.

f. Water leaving the box at high velocities may cause downstream scour at the streambed.

8-35. Inspection

a. Check for sag of the culvert floor. In times of light flow, this may be noted by location of sediment. Where there are several feet of water in the box, a profile of the crown may be taken.

b. Check for sag in the profile of the roadway overhead.

c. Check for vertical differential settlement at the expansion joints.

d. Check for transverse and longitudinal differential settlements at the expansion joints.

e. Check for widely opened expansion joints. Water may be seeping through joints from soil outside.

f. Check for canted wingwalls. This condition may be due to settlement, slides, or scour.

g. Check for slide failures in the fill around the box. Such slides are likely to affect the box as well.

h. Check for cracks and spalls in the top slab. Longitudinal cracks indicate shear or flexure problems; transverse cracks indicate differential settlement. Cracks in the sides may be from settlement or from extremely high earth pressures. Note the size, length, and location of the crack. Look for exposed or rusty rebars.

i. Where there is no bottom slab, look for undermining of side footings.

j. Check for undermining at the ends of the box and under the wings.

k. Examine the inside of the box for large cracks and debris. This may indicate the need for a debris rack. Check the inlet end of the culvert for debris. Note whether vegetation is obstructing the ends of the culvert.

l. If the culvert floor is visible, check it for abrasion and wear.

m. Note any other signs of deterioration of the concrete box, especially those which suggest design error or construction omissions.

CHAPTER 9

FINAL DOCUMENTATION

9-1. Annual (Army) and biannual (Air Force) inspection documentation

The results from these inspections are used to plan and coordinate preventative maintenance operations on the bridges. The inspection documentation should include, but not be limited to the following:

a. *Jobsite inspection documentation.* This documentation will generally be the bridge inspection notebook as discussed in chapter 7 or a copy of the suggested format as shown in appendix B.

b. *Photographs and sketches.* Photographs or sketches of problem areas and of item requiring maintenance should be included in the report.

c. *Cost estimation.* Cost estimation for each repaired item should be provided so that installation engineers and maintenance personnel are able to use it for their maintenance and budget planning.

d. *Remarks and recommendations.* A brief summary of the inspection findings should be made by the inspector. Problem areas, those requiring immediate attention, and those requiring additional attention from a structural engineer should be pointed out in this section.

9-2. Triennial (Army) and every third biannual (Air Force) bridge inspection documentation

The results from these inspections are used both to plan and coordinate preventative maintenance operations on the bridges and to insure their overall safety. The documentation should generally be prepared by a qualified engineer who participated in the inspection. The inspection documentation should include, but not be limited to the following:

a. *Paragraph 1: Introduction and Background.* This paragraph must contain all the pertinent information for an inspection of a bridge:

(1) The official request for conducting an inspection.

(2) Individual conducting the inspection (name, position, and qualifications).

(3) Criteria for an inspection (references and documentation).

b. *Paragraph 2: Load Classification Summary.* Load classification should be performed by a qualitied engineer.

c. *Paragraph 3: Bridge Repair Cost Summary.* The bridge repair costs in this paragraph are for the recommended maintenance and repairs which should be accomplished over the next 2 years. The estimated costs are based on the cost of materials and labor required to complete the job.

d. *Paragraph 4: Bridge Data.* This paragraph should provide the following information:

(1) *Bridge data.* Provide as follows:

(a) *Installation.*

(b) *Bridge number.* The official number assigned to bridge.

(c) *Date of inspection.*

(d) *Inspected by.*

(e) *Location.* Usually described by route number, county and log mile.

(f) *Design load.* The live loading for which the bridge was designed will be stated if it is known. A structure widened or otherwise altered so that different portions have different live load designs is to have each live load specified.

(g) *Military load classification.*

(h) *Date built.*

(i) *Traffic lanes.* State the number of traffic lanes.

(j) *Transverse section.* Include items noted below."

(k) *Roadway width.* This shall be the minimum clear width between curbs, railings, or other restrictions. On divided roadways, such as is found on freeways under overcrossings, the roadway width will be taken as the traveled way between shoulders, but the shoulders and median width will also be given.

(l) *Sidewalks.* If only one is present, the side shall be noted thus: "1 @ 5.0' (east)." Measurement is recorded to the nearest one-tenth of 1 foot. Sidewalks on both sides are noted thus: "2 @ 5.0'." If there are no sidewalks, note "None."

(m) *Clearances.* A clearance diagram should be made for each structure which restricts the vertical clearance over the highway, such as overcrossings, underpasses, and through truss bridges. The minimum number of vertical measurements shown on the diagram will be at each edge of the traveled way and the minimum vertical clearance within the traveled way. The report will state the minimum roadway clearance. This will include each roadway on a divided highway. When a structure is of a deck or point truss type so that no vertical obstruction is present, the vertical clearance shall be noted on the report as "unimpaired." Vertical measurements are to be made in feet and inches and any fraction of an inch will be dropped back to the nearest inch, i.e., a field measurement of 15 feet, 7 3/4 inches will be recorded as 15 feet, 7 inches. Horizontal measurements are to be recorded to the nearest one-tenth of 1 foot.

(n) Total length. This shall be the overall length to the nearest foot and shall be the length of roadway which is supported on the bridge structure. This will normally be the length from paving notch or between back faces of backwalls measured along centerline.

(o) Spans. The number of spans shall be listed in the same direction as the log mile. Spans crossing state highways will be normally listed from left to right looking ahead on route. Spans are noted as follows: e.g., "1 @ 24.0'" or "3 @ 45.0'" or "1 @ 36.0', 5 @ 50.0', 1 @ 36.0', 27.0', 12.0'." Span lengths shall be recorded to the nearest one-tenth of 1 foot and it shall be noted whether the measurement is center to center (c/c> or clear open distance (clr) between piers, bents, or abutments. Measurements shall be along centerline.

(p) Skew. The skew angle is the angle between the centerline of a pier and a line normal to the roadway centerline. Normally, the skew angle will be taken from the plans, and it is to be recorded to the nearest degree. If no plans are available, the angle is to be estimated. If the skew angle is 0 feet, it should be so stated.

(q) Plans available. State what plans are available, where they are filed, and if they are "as built."

(r) Inspection records. Record the year inspected, inspector, and qualification. Under description, briefly give all pertinent data concerning the type of structure. The type of superstructure will generally be given first followed by the type of piers and type of abutments along with their foundation. If the bridge is on piles, the type of piles should be stated. If data are available, indicate type of soil upon which footings are founded, maximum bearing pressures, and pile capacities.

(2) Bridge component rating. Each component of the bridge should be numerically rated. A suggested format and rating system are provided in appendix C.

(3) Recommendations and repair costs. The final portion of appendix C provides a space for recommendations regarding deficient bridge components.

(4) Sketches.

(a) Overall sketch. The first sketch schematically portrays the general layout of the bridge, illustrating the structure plan and elevation data. The immediate area, the stream or terrain obstacle layout, major utilities, and any other pertinent details should also be included.

(b) Bridge component sketches. These are necessary for a component condition rating of "6" or "7" (refer to appendix C).

(c) Substructures. Sketches or drawings of each substructure unit should be included. In many cases it will be sufficient to draw typical units which identify the principal elements of the substructure. Each of the elements of a substructure unit should be numbered so that they can be cross referenced to the information appearing on the data page on the left-hand side of the sketch. Items to be numbered include piling, footings, vertical supports, lateral bracing of members and caps.

(d) Special sketches. Additional sketches may have to be prepared of critical areas of certain bridges.

(5) Photos. At least two photographs of each bridge, one showing a roadway view and one showing a side elevation view, should be provided. Other photos necessary to show major defects or other important special features also may be included. A photo showing utilities on the structure is desirable. All signs of distress, failure, or defects worthy of mention, as well as description of condition and appraisal, should be noted with sufficient accuracy so that another inspector at a future date can easily make a comparison of condition or rate of disintegration. Photographs and sketches should be used freely as needed to illustrate and clarify conditions of structural elements. Good photos and pictures are very helpful at future investigations in determining progression of defects and to help determine changes and their magnitude. All recommendations and directions for corresponding repair and maintenance should be included.

CHAPTER 10

GENERAL PREVENTIVE MAINTENANCE, REPAIR, AND UPGRADE

Section I. INTRODUCTION

10-1. General

With the cost of constructing and replacing bridges escalating every day, it is imperative that we make the most out of our existing bridges. The formula for doing this is: properly maintaining each bridge to extend its service life, immediately repairing any structural damage or deterioration of the bridge to prevent increased damage or deterioration, and upgrading the load capacity of the structure to meet the future increased traffic requirements. The specific categories of bridge maintenance, repair, and upgrade are discussed in the following paragraphs.

10-2. Preventive maintenance

Maintenance is the recurrent day-to-day, periodic, or scheduled work that is required to preserve or restore a bridge to such a condition that it can be effectively utilized for its designed purpose. It includes work undertaken to prevent damage to or deterioration of a bridge that otherwise 'would be more costly to restore. The concept of preventive maintenance involves repair of small or potential problems in a timely manner so that they will not develop into expensive bridge replacements. Preventive maintenance activities can be divided into two groups: those performed at specified intervals and those performed as needed.

   a. Specified interval maintenance. This group includes the systematic servicing of bridges on a scheduled basis. The interval varies according to the type of work or activity. Tasks identified as interval maintenance can be incorporated into a maintenance schedule for that bridge. Examples are:

   (1) Cleaning drainage facilities.
   (2) Cleaning and resealing expansion joints.
   (3) Cleaning expansion bearing assemblies.

   b. As-needed maintenance. These activities are performed when the need is foreseen for remedial work to prevent further deterioration or the development of defects. The need for this type of maintenance is often related to the environment or identified during inspections. Example activities include:

   (1) Sealing concrete decks.
   (2) Painting steel members.
   (3) Snow and ice removal.

10-3. Replacement

The replacement of bridge member components is based on the material of the existing member, equipment availability, and the training level of the repair crews. More detailed considerations for the replacement of each type of bridge member are provided in section III of chapters 11 through 13.

10-4. Repair

Bridge repair is actually an extension of a good maintenance program. It involves maintaining the bridge's current structural load classification. Selection of the correct repair technique for a bridge of any type and material depends upon knowing the cause of a deficiency and not its symptoms. If the cause of a deficiency is understood, it is more likely that the correct repair method will be selected and that the repair will be successful. A general procedure to follow for designing and executing a repair involves the evaluation and determination of the causes for the deficiency and the methods, materials, and plans to be used in the execution of the repair.

   a. Evaluation. The first step is to evaluate the current condition of the structure. Items to include in the evaluation are:

   (1) Review of design and construction documents.
   (2) Review of structural instrumentation data.
   (3) Review of past bridge inspections.
   (4) Visual examination, nondestructive test, and laboratory test.

   b. Relate observations to causes. Evaluation information must be related to the mechanism or mechanisms that caused the damage. Since many deficiencies are caused by more than one mechanism, a basic understanding of the causes of deterioration is needed to determine the actual damage causing mechanism.

   c. Select methods and materials. Once the underlying cause of the structural damage is determined, selection of appropriate repair materials and methods should be based on these considerations:

   (1) Determine prerepair adjustments or modifications required to remedy the cause, such as changing the water drainage pattern, correcting

differential foundation subsidence, and eliminating causes of cavitation damage.

(2) Determine constraints such as access to the structure, the operating schedule of the structure, and weather.

(3) Determine permanent/temporary repair advantages/disadvantages.

(4) Determine the available repair materials and methods and the technical feasibility of using them.

(5) Determine the most economically viable, technically feasible methods and materials. Select the combination that ensures a satisfactory job.

d. Prepare design memoranda, plans, and specifications. This step should be based on existing guide specifications, enhanced by incorporating experience gained from similar projects, and allowing as much flexibility with regard to materials as possible.

e. Execute the repair. The success of the repair depends on the degree to which the repair is executed in conformance with the plans and specifications.

10-5. Bridge upgrade

a. General. The upgrading of existing bridges is usually required where they are to carry heavier live loads than those for which they were designed. Upgrading or strengthening may also be required because of inadequate design or as the result of localized deterioration. The decision to upgrade a bridge or to replace it should take into account the age of the structure, the material of which the various members are made, the fatigue effect of the live loading, the comparative estimated cost, the added service life of the upgraded bridge, and the possible future increase in the live loading.

b. Upgrade levels. The upgrade of a bridge structure can be carried out at three levels.

(1) Strengthening of the existing individual components of the bridge to provide a moderate increase of the bridge's load carrying capacity.

(2) Redesigning the structure by adding components (stringers, piers, load bearing decks, etc.).

(3) Redesigning the structure by a combination of strengthening existing components and adding components to increase the load capacity.

Section II. COMMON MAINTENANCE TASKS

10-6. General

There are numerous different types of bridges and materials of which these bridges are constructed. However, there are some maintenance tasks that are common to all bridges despite their individual designs and construction materials. These tasks are incorporated into standardized maintenance operating procedures and generally involve keeping the bridge clean and conducting work and minor repairs to prevent bridge deterioration.

10-7. Cleaning deck drains

Drains and scuppers should be open and clear to ensure that the deck drains properly and that water does not pond. Ponding of water on the deck increases the dead load on the bridge and presents a hazard to drivers in the form hydroplaning. Proper drainage also helps prevent water from leaking through the deck or deck joints and causing deterioration of other superstructure components.

10-8. Ice and snow removal

The primary reason for the removal of snow and ice is to provide a safe bridge for motorists. Bridges are generally the first portion of the road network to ice over and require immediate attention in freezing weather. The primary means to combat the accumulation of snow and ice is plowing the snow from the traffic lane of the bridge, spreading abrasives (crushed rock, sand, cinders, etc.) to improve the wheel traction, and chemicals (see table 10-1) to lower the freezing point of the water on the deck. When deicing salts (calcium chloride or sodium chloride) are used as part of this process, it is imperative that the maintenance schedule includes cleaning the bridge in the spring to remove any lasting effects of the salts. Any abrasives used on the structure should be removed as soon as possible after the snow period is over to reduce wear on the deck.

10-9. Bank restoration

Bank restoration involves the area in and around the abutments and up to the waterline. Erosion is the biggest problem and a maintenance program should include filling in washouts and seeding or using riprap to help prevent erosion. For a more detailed description on the use of fill and riprap, refer to paragraph 10-18d.

10-10. Traffic control items

It is important that traffic control items (clearances, load classifications, speed signs, centerlines, etc.) be maintained on a regular basis to control the traffic across the bridge. It is especially impor-

Table 10-1. Chemical application rates

Temperature	Conditions	Precipitation	Multilane Divided	Primary	Secondary	Instructions
30 °F and above	Wet	Snow	.300 salt	300 salt	300 salt	- Wait at least 0.5 hr before plowing
		Sleet or freezing rain	200 salt	200 salt	200 salt	- Reapply as necessary
25-30 °F	Wet	Snow or sleet	Initial at 400 salt repeat at 200 salt	Initial at 400 salt repeat at 200 salt	Initial at 400 salt repeat at 200 salt	- Wait at least 0.5 hr before plowing; repeat
		Freezing rain	Initial at 300 salt repeat at 200 salt	Initial at 300 salt repeat at 200 salt	Initial at 300 salt repeat at 200 salt	- Repeat as necessary
20-25 °F	Wet	Snow or sleet	Initial at 500 salt repeat at 250 salt	Initial at 500 salt repeat at 250 salt	1,200 of 5:1 sand/salt; repeat same	- Wait about 3/4 hr before plowing; repeat
		Freezing rain	Initial at 400 salt repeat at 300 salt	Initial at 400 salt repeat at 300 salt		- Repeat as necessary
15-20 °F	Dry	Dry snow	Plow	Plow	Plow	- Treat hazardous areas with 1,200 of 20:1 sand/salt
	Wet	Wet snow or sleet	500 of 3:1 salt/calcium chloride	500 of 3:1 salt/calcium chloride	1,200 of 5:1 sand	- Wait about 1 hr for plowing; continue plowing until storm ends; then repeat application
Below 15 °F	Dry	Dry snow	Plow	Plow	Plow	- Treat hazardous areas with 1,200 of 20:1 sand/salt

note: Chemical application is in pounds per mile of 2-lane road or 2 lanes of divided.

tant for moveable bridges that navigation lights, traffic control systems, and protective fender systems be monitored regularly. This is a safety more than a structural issue; however, it constitutes an important component in providing a complete maintenance program.

10-11. Bearings and rollers

All rockers, pins, and rollers are to be kept free of debris and corrosion, lubricated where necessary, and maintained in good working order. Depending on the type of bearing (fixed or expansion), they should permit the superstructure to undergo necessary movements without developing harmful overstresses. A "frozen" or locked bearing that becomes incapable of movement allows the stresses generated to become excessive and may even cause a major failure in some affected member.

10-12. Debris and removal

a. Superstructure. Any debris left on the superstructure due to traffic or high water should be removed for safety reasons and to prevent deterioration in areas were the debris will trap moisture onto the superstructure.

b. Substructure. The substructure is susceptible to debris or floating ice that forms drifts against its components. This can cause premature deterioration of these components and place excessive lateral loads on the whole structure. The techniques available to remove drifts are:

(1) Clear small debris with a pole or hook.

(2) Pull large pieces of debris clear with a crane.

(3) Clear large and small pieces of debris with a powerboat.

(4) Blast large jams to break them up.

10-13. Bridge joint systems

a. General. Joints are designed to provide for rotation, translation, and transverse movements of the superstructure under live loading and thermal expansion. The system should also prevent water leakage onto the components below the bridge deck. The two common joint systems are open and closed joints.

b. Open joints. The open joint provides for longitudinal movement of the superstructure. The joint construction is not watertight and should permit traffic to cross smoothly.

(1) *Finger joints.* Interlocking steel fingers attached to a steel plate allow longitudinal but restrict transverse deck movements.

(a) *Clogged joint and drain trough.* Frequently flush and clean the joint and drainage system to remove debris accumulation in the system. This will also help prevent corrosion and concrete deterioration.

(b) *Loose joints.* Remove loose or faulty bolts or rivets, reposition the expansion device, and rebolt. It may be necessary to countersink the bolts or rivets to avoid future problems.

(c) *Broken finger joints.* Weld replacement fingers onto the joint.

(d) Fingers closed. Trim the expansion fingers or remove the system, reposition, and reinstall.

(2) Armored joints. These consist of steel angles at concrete edges which are left open or filled with a mastic or other material to prevent intrusion of debris. If the joints are clogged, clean out the joint, repair any broken angles, and apply a liquid polyurethane or preformed compression joint sealant for waterproofing and to prevent debris intrusion.

(3) Sliding plate. A horizontally positioned steel plate is anchored to the deck and allowed to slide across an angle anchored to the opposite face of the opening.

(a) Clogged expansion gap. Remove any dirt, debris, or asphalt from the gap to ensure that sliding plate interacts properly with its angle seat.

(b) Joint closed. Trim the steel plate.

c. Closed joints. The closed joint is a watertight arrangement of various materials which allow longitudinal movement of the superstructure.

(1) Elastomeric. A sealed, waterproof joint system which uses steel plates and angles molded into neoprene coverings to provide an anchorage and load transfer (figure 10-1).

(a) Faulty section. Remove and replace.

(b) Inadequate seal. Apply new sealant or remove and reinstall using proper sealing techniques.

(c) Loose or broken bolts. Remove broken bolts and replace with "J" bolts.

(2) Compression seal. Extruded neoprene with cross-sectional design and elasticity to provide for retention of its original shape. Leakage is the most

common failure associated with this joint sealant and requires replacement of the deficient compression joint sealant (figure 10-1).

10-14. Scour protection

The excavation or removal of the soil foundation from beneath the substructure undermines the load carrying capacity of the bridge. It can also cause excessive settlement if proper preventive maintenance is not practiced. A foundation may be protected against scour in the initial design and after construction has been completed.

a. Design.

(1) Locate bents and piers parallel to the direction of the water flow.

(2) Use long piles driven to a depth which provides sufficient bearing and accounts for scouring.

(3) Drive a row of closely spaced fender piles perpendicular to the stream flow at the upstream end of the substructure. This may not protect the downstream end since eddies and increased velocity may produce erosion on the downstream end. Their greatest effectiveness is for narrow piers and larger spans structures.

b. After construction.

(1) Place sandbags around the base of bents, piers, and abutments, particularly at the upstream end.

(2) Place riprap consisting of stones weighing at least 50 pounds or bags filled with stones or cement.

(3) Divert drainage lines when scouring is due to local ground drainage or drainage from the deck itself.

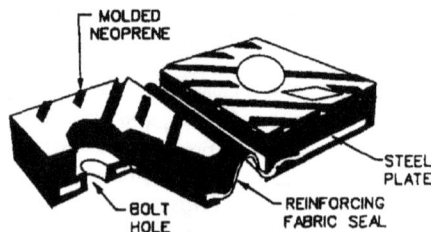

MODEL INCORPORATING A GLAND TYPE OF JOINT SEALANT.

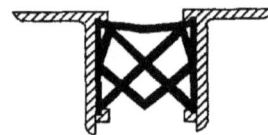

COMPRESSION JOINT SEAL MODIFIED WITH CORNER CLIPS

Figure 10-1. Examples of closed joints.

Section III. COMMON REPAIR TASKS

10-15. General

Just as in maintenance tasks there are several repair tasks that apply to almost all bridges despite the materials used in construction. The purpose of bridge maintenance is to bring a bridge

back up to original design load after damage or deterioration to the structure. These repairs can involve the strengthening, replacing, or adding support to the existing components of the structure. In the case of common repair tasks, these tasks generally involve repair to the foundation or substructure of the bridge.

10-16. Abutment stability

a. In addition to providing end support for the bridge deck, an abutment also acts as a retaining wall and is subject to horizontal earth pressures. These pressures coupled with the dynamic loading of vehicle traffic has the tendency to push out the abutment. If the abutment is unstable, it may be shored or fixed using guylines from shore anchorages or a deadman tie-back system.

b. The procedure for this process is as follows (figure 10-2): Emplace a deadman or drive a pile anchor approximately 3 feet on either side of the approach to the bridge. These anchors should be from 60 to 100 feet from the face of the existing abutment. Drill a hole in the wing-wall on both sides of the abutment and in a position outside and in line with the abutment cap. Run a restraining rod or cable from the deadman through the hole in the wall. Place a beam (example: steel wafer) on the outside of the cap. The beam must extend a distance at least greater than the position of the drilled holes in the wingwall on both sides of the abutment cap. Connect the retaining rod or cable to the beam and place tension on the rods or cables. This technique can be used on smaller abutments to help draw the abutment back to its original position and to hold it in place.

10-17. Drift and floating ice

Drift and floating ice place forces on the piers and frame bent of bridges. They can also cause structural damage to the cross section of the piles or columns of the system from the impact of the debris on the substructure. A common repair to minimize this problem is to install dolphins and/or fender systems upstream of the piers to adsorb the energy of the physical contact with the drift and to help break it up.

a. Pile cluster dolphins. This type of dolphin consists of a cluster of piles with the tops pulled together and fixed into position (figure 10-3). Dolphins can be constructed from:

(1) *Timber.* The top is lashed together with wire rope. Timber piles can be protected from impact damage by banding the piles with sheet metal.

(2) *Steel tube.* Dolphin system is connected with bracing and fender arrangement.

(3) *Caisson.* Sheet-pile cylinders of large diameter filled with sand or concrete and topped with a concrete slab. Fendering can also be attached to the outside sheets, if needed.

b. Fenders. Fenders are designed to divert the flow of drift and ice around the piers. These fenders generally form a wedge on the upstream side of the pier to divert the flow.

(1) *Timber/steel bents.* A series of piles with timber wales and braces are attached to the top portion of the bent system. The steel piles may be tied together using a concrete slab (figure 3-19).

(2) *Steel or concrete frames.* Steel or concrete frames are sometimes cantilevered from the pier and faced with timber or rubber cushioning to reduce collision impact from surface craft and debris.

(3) *Timber grids.* Timber grids, consisting of post and wales, are attached directly to the pier (figure 3-19).

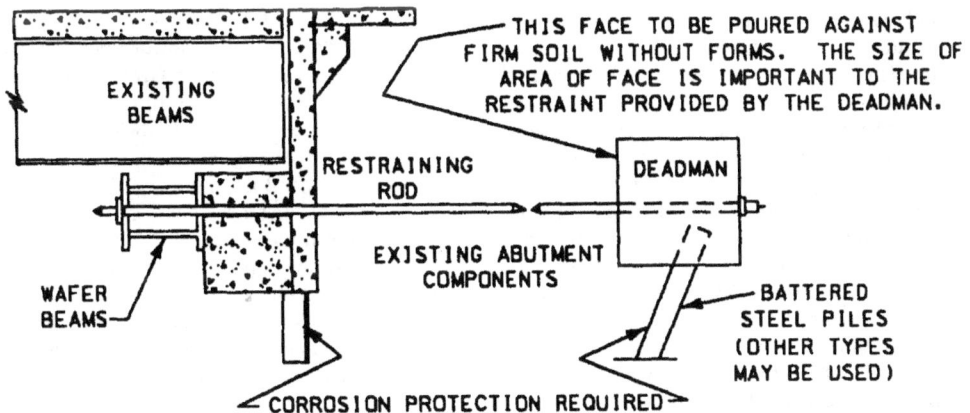

Figure 10-2. Abutment held in place with a deadman.

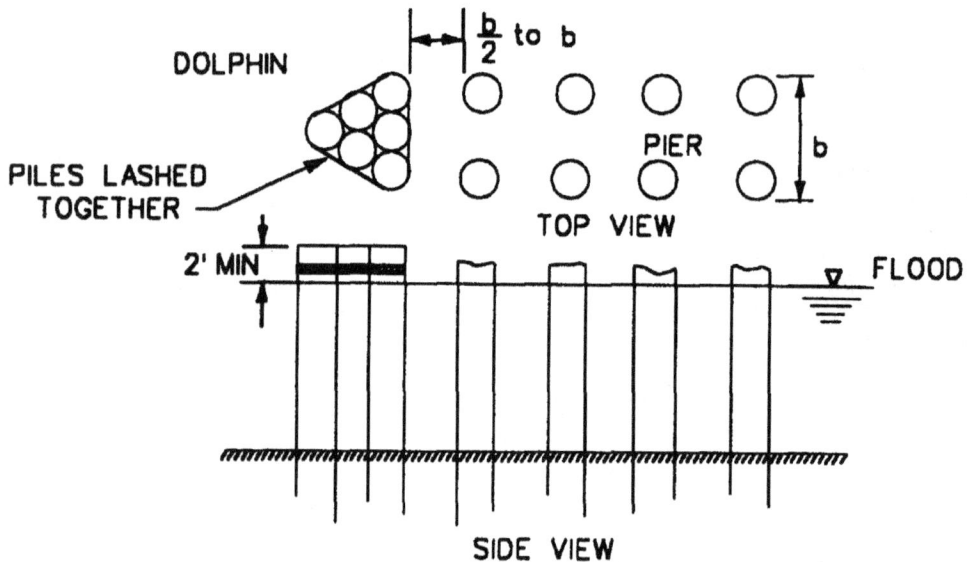

Figure 10-3. Typical use of dolphins.

10-18. Scour

When scour undermines the existing foundation, methods must be undertaken to reestablish the foundation. Scour can effectively reduce the bearing of piles, undermine pier footings and abutments, and cut into the bank.

a. *Piles.* When scour reduces the effective bearing of the piles in a pier:

(1) Additional piles can be added to the base of the pier to make up for the lost bearing and riprap added to prevent future scouring (figure 12-12).

(2) A concrete footing can be added to the base of the pier to make up for the lost bearing as follows (figure 10-4). Place a tremie encasement around the bottom of the pier. Inject concrete or mortar into the encasement. The concrete will displace the water from within the encasement. The formwork or encasement can be removed after the concrete is cured. Nails or spikes can be driven into timber piles, shear studs or bolts placed on steel piles, and rebar placed in drilled holes or the outer surface chipped on concrete piles to provide a better bond between the pile and the footing.

b. *Pier footings.* When footings are undermined, the most common repair method is to fill the void foundation area with a concrete grout or crushed stone. To place grout, some type of formwork must be used to confine the grout.

(1) *Concrete grout.*

(a) *Tremie encasement.* This is a steel, wood, or concrete form that is placed around the existing footing to reestablish the foundation. The form

allows the concrete grout to be pumped under the eroded footing and displaces the water in the encasement through vents (figure 10-4).

(b) *Confinement walls.* Walls of stone, sandbags, or bags filled with riprap are placed along the faces of the footing and extending through the mud layer of the river bottom. The grout is injected into the cavity below the footing and the water is displaced through the voids in the wall. The grout penetrates into the voids in the wall and seals the confinement wall. The cavity is filled with grout and the foundation is reestablished (figure 10-5).

(c) *Flexible fabric.* A closed bag of canvas, nylon, etc., with grout injection ports is positioned under the footing. Grout is pumped into the bag and it expands to fill the cavity. The injection port is then closed and fabric confines the grout until it can cure (figure 10-5).

(2) *Backfill.*

(a) *Dry footing.* In this case the footing is not under water and may have been eroded by service runoff or flooding. A good structural fill material can be compacted into the erosion cavity to fill the void. If the streambed is eroded below the base of the footing, the compacted fill will be extended on a slope of 2 to 1 from the current streambed to the base of the footing. Riprap will then be placed around the footing to prevent further scouring.

(b) *Wet footing.* Crushed stone is used as the fill material (figure 10-6).

c. *Abutments.* Scour around the base of abutments can be repaired in a similar fashion as the pier footing as follows (figure 10-7):

(1) Shore up the abutment to prevent settlement during the repair,

(2) Remove any loose material from the scoured area,

(3) Set bolts into the abutment face along the length of the abutment. These bolts should extend 3 to 6 inches from the abutment face and be spaced 2 feet apart. Use the bolts installed in step 3 to connect an expansion shield to the abutment. Place concrete behind the shield to fill the erosion cavity and the space between the shield and abutment face. Place riprap on a 2 to 1 slope to prevent future scouring.

d. *Bank slope.* Erosion under and around a concrete slope protector can be repaired using a riprap till or may require extending the protector.

(1) *Riprap* (figure 10-8). Fill the scour hole with riprap. Extend the riprap above the face of the concrete to protect from future scouring. Slope the riprap level down from the edge of the protector to the face of the concrete protector.

(2) *Protector extension (figure 10-9).* Remove loose material from the scour hole. Backfill with sand or gravel. Form a ground mold in the backfill for the extended slope protector. Place concrete into the ground mold. Add the following steps for an undermine protector: cut a hole in the protector above the erosion cavity, backfill or grout through this hole, and repair the holes.

10-19. Settlement

Foundation settlement usually is caused by structural failure of the foundation material or scour.

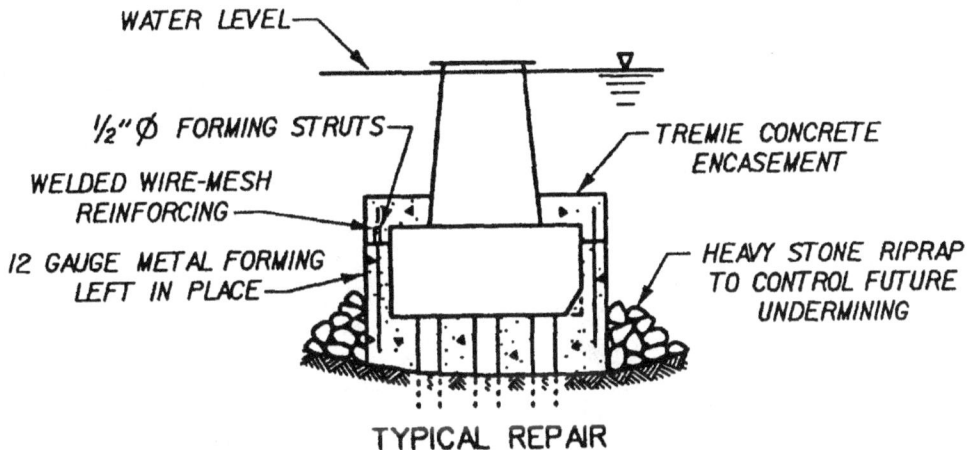

Figure 10-4. Forming a footing with a tremie encasement.

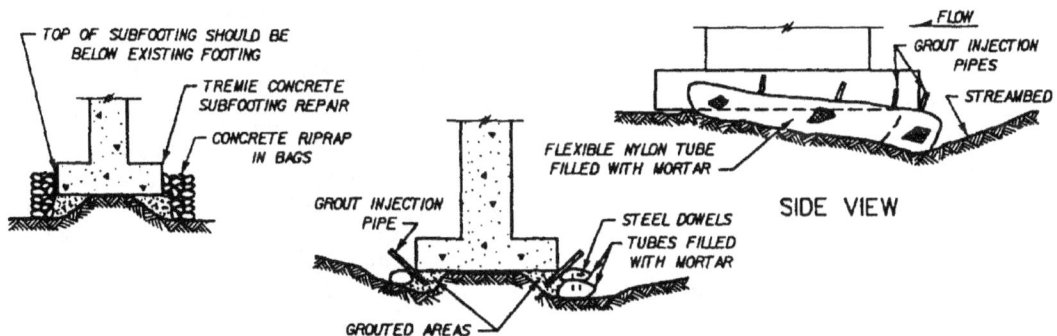

Figure 10-5. Alternate methods for confining grout under footings.

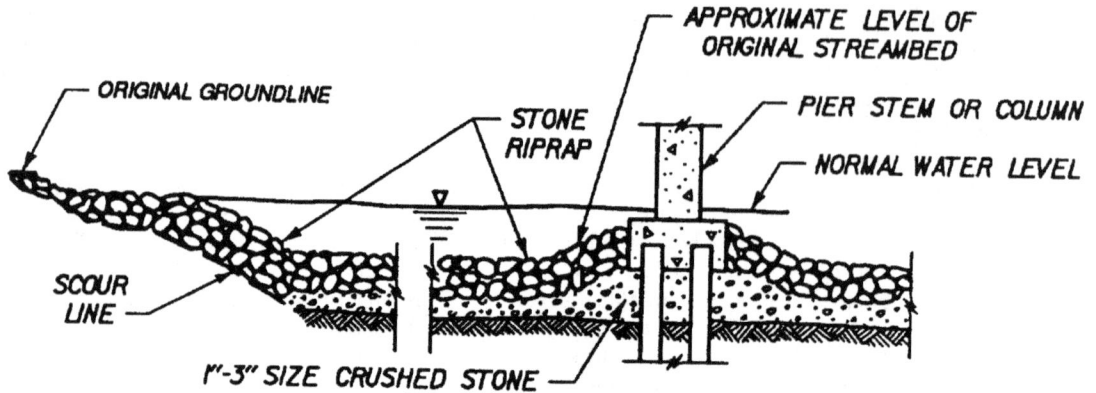

NOTE: THE USE OF THIS SIZE CRUSHED STONE MAY BE PROHIBITIVE IF STREAM CURRENTS ARE STRONG

Figure 10-6. Use of crushed or structural fill to repair scour damage.

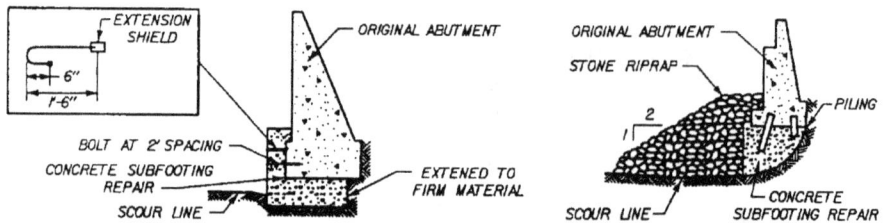

Figure 10-7. Repair of scour around concrete abutments.

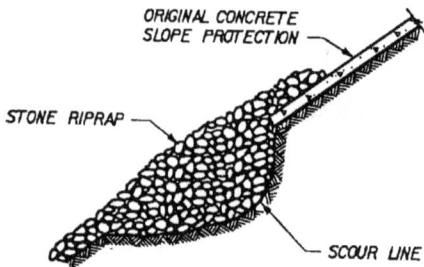

Figure 10-8. Bank repair using riprap,

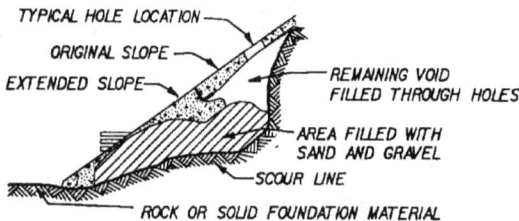

Figure 10-9. Concrete bank protector extension.

a. Minor settlement. This can be corrected by jacking up the structure and inserting steel shims between stringers and cap or between bearing plates and pedestal. Hard wood shims can be used under wooden members.

b. Major settlement. The pier is tilted or has settled to a point that it can no longer safely carry its design load.

(1) *Piles.* Correct the cause of the settlement by adding piles or a footing (Refer to related paragraphs in the following chapters for specific methods). Jack the deck to its proper position. Enlarge or replace the existing cap. Level the deck and remove jacks.

(2) *Pier footings.* Correct the cause of the settlement. Jack the deck to its proper position. Construct a level bearing surface for the superstructure. For bents, enlarge or replace the existing cap. For walls, enlarge the stem of the pier by placing a form around the stem and placing concrete. The concrete formwork should extend to the top of the pier to provide a level bearing surface (figure 10-10). Then cure, level the deck, and remove the jacks.

10-20. Waterway

Some waterways with flat gradients and flood plains have a tendency to shift channel locations.

Such channel shifts may deposit eroded material: against the bridge piers, erode pier foundations, or attack the approach. This can be controlled by dikes of earth, rock, or brush mats.

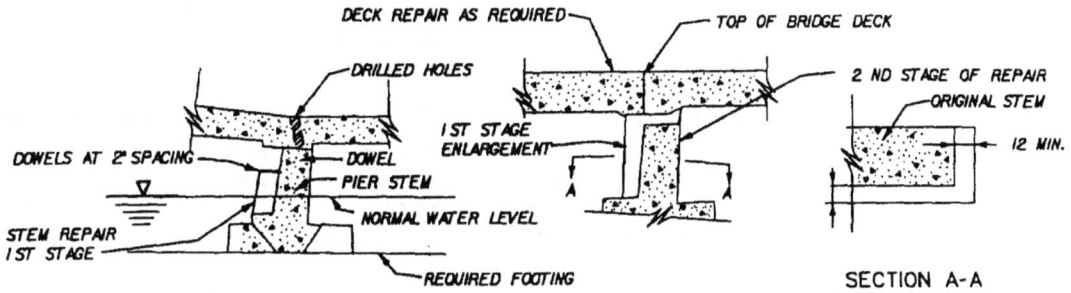

(a) SECTION THRU PIER STEM SHOWING FIRST STAGE OF REPAIR

(b) SECTION THRU STEM SHOWING SECOND STAGE OF REPAIR

SECTION A-A

NOTE: THIS REPAIR SHOULD BE UNDERTAKEN ONLY IF THE STABILITY OF THE PIER AND OTHER ATTENDANT DAMAGES ARE JUDGED TO BE ACCEPTABLE AND AMENABLE TO REPAIR

Figure 10-10. Settlement repair of a concrete wall pier.

Section IV. COMMON METHODS TO UPGRADE EXISTING BRIDGES

10-21. General

There are several methods available to upgrade existing bridges. Some of the more common methods involve shortening the effective span length, adding stringers, strengthening the piers, reducing the dead load, and strengthening individual structural members. This section will discuss a few of the techniques involved with these upgrade methods.

10-22. Shortened span lengths

a. *Bearing point.* Shifting the bearing point can achieve a slight increase in bridge strength and can be used in conjunction with other upgrade methods to obtain a total bridge upgrade. The bearing point is shifted in toward the span, and the strength increase is limited by the size of the bearing surface on the abutment or pier cap. The shift can be accomplished by:

(1) Jacking the stringer off the bearing point and moving the existing point out on the bearing face.

(2) Placing a new bearing point in front of the existing point and using wedges to transfer the load to the new point.

b. *Intermediate piers.* The greatest bridge strength can be achieved by adding intermediate piers between the existing piers and abutments. The original span acts as a continuous beam and the negative movement must be checked over the intermediate pier.

c. *A-Frame.* For short spans, the A-frame provides an expedient substitute for installing an intermediate pier. The legs of the frame are anchored to the existing pier footings or footings are constructed for this purpose. These footings must be designed or reinforced to carry the lateral thrust transmitted through the frame's legs. The apex of the A-frame supports a cap which has a bearing surface for the bridge stringers. The use of the A-frame reduces the clearance below the bridge and may require additional horizontal reinforcement against seasonal floods (figure 10-11, part a).

d. *Knee-braces.* Knee-braces may be used to

function in nearly the same manner as A-frames. However, the reaction of the span is not that of a continuous span and must be analyzed as support bracing (figure 10-11, part *b).*

10-23. Add stringers

The addition of stringers may require respacing existing stringers. The procedure increases the capacity of the existing deck and redistributes the loads carried by the stringers. When the deck is not replaced, special techniques must be used to place the new stringers. With the increase of the load carrying capacity of the superstructure, it is imperative to check the substructure for the same loading conditions.

10-24. Strengthen piers

a. Add piles. Drive additional piles and tie into the existing pier system.

b. Add or increase a footing. Once the footing is established, columns can be added between the new footing and the existing cap to provide a greater column strength in the pier.

c. Add or enlarge columns. The load carrying capacity of a pier or bent system can be increased by adding more columns or by adding material to the existing column cross sections.

10-25. Reduce deadload

The live load capacity of a bridge can be increased by reducing the applied dead load, thereby allow-ing existing capacity to be used for carrying increased live load. Significant dead load reductions can be obtained by removing an existing concrete deck and replacing it with a lighter deck. A reinforced concrete deck 8 inches thick weighs approximately 100 pounds per square foot. This weight is compared to the weights of lightweight decks in table 10-2.

10-26. Posttensioned bridge components

Posttensioning can be used to:

a. Relieve tension, shear, bending, and torsion overstress.

b. Add ultimate strength to the bridge structure.

c. Reverse displacements.

d. Change simple span to continuous span behavior. Various methods and configurations for posttensioning are shown in table 10-3.

10-27. Strengthen individual members

In many cases the bridge classification can be increased by strengthening the individual members of the structure. The drawbacks to this method are that it generally increases the dead load of the structure and the added material shall be considered effective in carrying the added dead load and live load only. Methods for strengthening individual members are dependent upon the material type and are thus discussed in the following chapters.

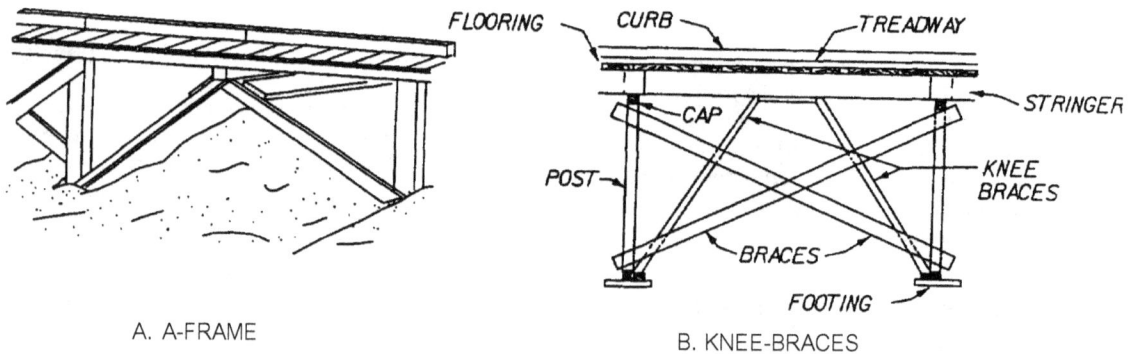

Figure 10-11. Expedient methods of span length reduction.

Table 10-2. Lightweight decks

Lightweight Deck	Description	lb/sq ft
Open-grid steel deck	Steel stringers connected by a welded steel plate grid that provides the deck surface. Spans 5-9 feet.	15-25
Concrete-filled steel grid	Half-filled: S-inch deck with concrete placed in the top half of the grid. Full-depth fill: S-inch deck with concrete placed full depth in the grid.	46-51 76-81
Exodermic deck	A 3-inch prefabricated concrete slab joined to a steel grating. Spans 16 feet.	40-60
Laminated Timber deck Transverse Orientation	Vertically laminated 2-inch-diameter lumber bonded together into structral panels 48 in. wide with a depth from 3 1/8 to 6 3/4 in. Longitudinal Orientation	10.4-22.5
Lightweight concrete deck	Concrete mix uses lightweight aggregate (expanded shale, slate, or clay). These decks can be cast in place or factory precast.	75
Orthotropic plate deck	Aluminum alloy plates with a skid-resistant polymer wear surface reinforced by extrusions. The deck bolts to the existing beams using hold-down brackets. Steel plates - No standard design.	20-25 45-130

Table 10-3. Bridge posttensioning configurations

Method	Beam/Stringer Configuration	Truss Configuration
Eccentric Tendon Includes axial compression and negative bending		
Concentric Tendon Relieves tension members by applying axial compression.		CONCENTRIC TENDONS ON INDIVIDUAL MEMBERS CONCENTRIC TENDON ON A SERIES OF MEMBERS
Polygonal Tendon Induces axial compression, nonuniform negative bending in the posttensioned region, and shear opposite to load shear.		
Polygonal Tendon with a Compression Strut Similar to polygonal but doesn't add axial compression into the existing structure.		
King Post Combination of eccentric and polygonal tendon configurations. The moment to axial force ratio is large due to the king post. Queen Post-uses 2 post.		
External Stirrup Timber - Sect. 13-7c(2) Concrete - Sect. 14-72(1) Box beam - Sect. 14-7e(2)	CONCRETE BEAMS EXTERNAL STIRRUP	BOX BEAMS EXTERNAL STRAP CROSS SECTION EXTERNAL LATERAL TIE

CHAPTER 11

STEEL BRIDGE MAINTENANCE, REPAIR, AND UPGRADE

Section I. PREVENTIVE MAINTENANCE FOR CORROSION

11-1. General

Rust and corrosion are the greatest enemies of steel. When rust is allowed to progress without interruption, it may cause a disintegration and subsequently complete loss of strength in a bridge member. The corrosion also causes other problems such as pressure or friction between the surfaces.

11-2. Structural steel

Preventive maintenance of steel bridge components consists mainly of measures to protect the steel from corrosion. The preservation of steel involves protection from exposure to electrolytes, such as water or soil. When deicing salt is added to the electrolyte, there is a dramatic increase in the rate of corrosion of the structural steel. Corrosion is usually easily spotted by visual inspection. Protection from corrosion can take various forms:

a. *Weathering steel.* This special type of steel forms its own protective coating and theoretically does not need painting. However, many state highway departments have indicated poor performance from their bridges constructed with this type of steel. Therefore, members constructed from weathering steel should be monitored for excessive corrosion and painted if necessary.

b. *Paint.* Typical painting requirements are based on whether the steel is new or is to be repainted. The following steps are usually necessary:

(1) *New steel.* One prime coat applied in the shop, one prime coat applied in the field, two color coats applied in the field.

(2) *Repainting (depends on the condition of the existing paint).* If cleaned to bare metal, use one prime and two color coats. If cleaned to prime coat, use two color coats. If no prime exposed, use one color coat.

(3) *Paint seal.* The intersections and edges of metal surfaces can be protected from corrosion with a paint seal. This is a paste paint that prevents moisture penetration between the metal parts.

(4) *Notes.*

(a) When removing lead-based paint, precautions must be taken to protect against lead inhalation, ingestion, and pollution.

(b) Schedule field painting at the end of maintenance projects to avoid damage to the fresh paint.

(c) Separate cleaning and painting operations to avoid contaminating the fresh paint.

(d) Protect cleaned steel until paint can be applied.

(e) Apply paint to a dry surface that is not too hot. This ensures a good bond that does not blister (ambient temperature greater than 40 °F, relative humidity less than 85 percent).

c. *Cathodic protection.* Zinc or aluminum anodes are attached to H piles to abate corrosion of steel in salt or brackish water (figure 11-1). Small zinc anodes are used when less than 8 linear feet of pile is exposed. Large zinc or aluminum anodes are used when greater than 8 linear feet of the pile is exposed.

d. *Good housekeeping.* Steps to follow:

(1) Keep drains open to remove standing water from steel surfaces.

(2) Keep deck joints watertight to prevent water leakage onto steel members.

(3) Keep exposed areas clean by pressure washing.

(4) Spot paint and repaint as necessary.

(5) Maintain steel cables by removing foreign objects from the cable support system, cleaning and lubricating cable supports, tightening and replacing stirrups, and repairing cable wrapping.

Section II. REPAIR AND STRENGTHEN

11-3. General

a. *Repair decision.* Each repair decision must carefully weigh the long-term operational requirements and existing environmental factors that can help accelerate corrosion prior to evaluating initial

and life cycle costs. The physical condition of the structure must first be determined by a detailed inspection. The structural capacity of the steel should be known. Once the physical condition of the bridge is evaluated, a determination of

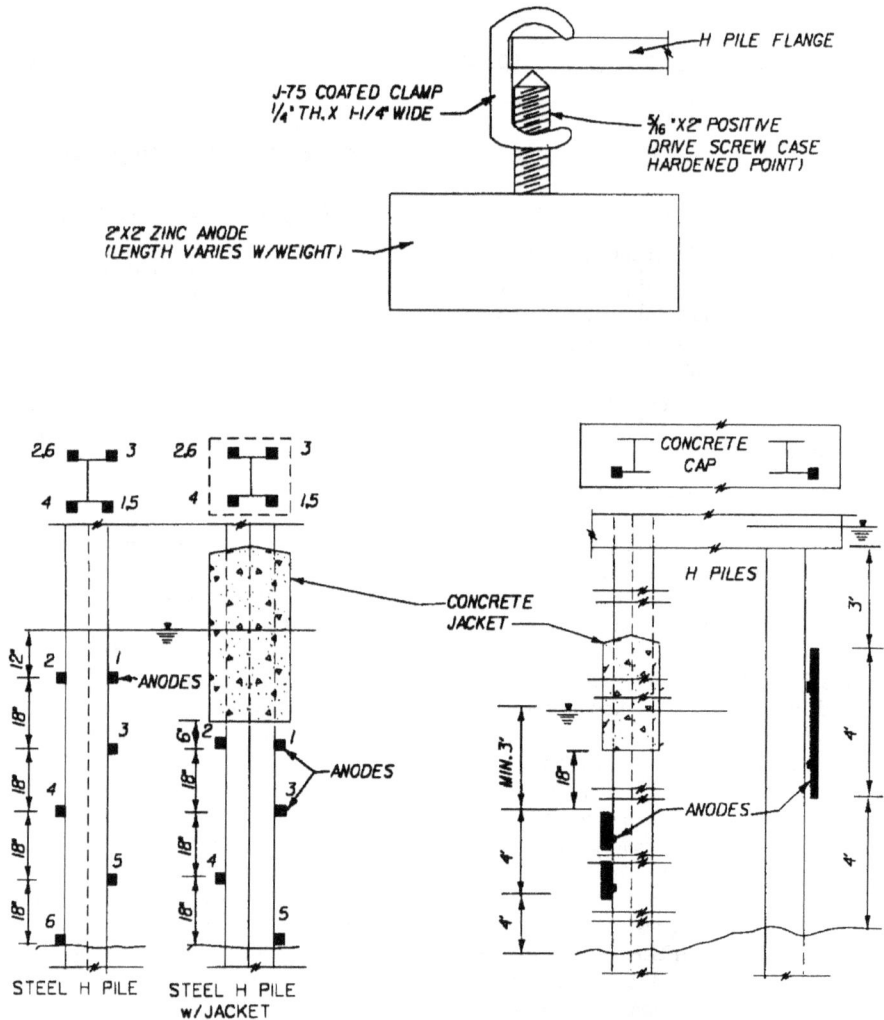

Figure 11-1. Anodes placed on steel H piles for corrosion protection

whether damaged bridge components should be repaired or replaced is made.

b. Common repairs. The most common steel repairs are:

(1) Adding metal to strengthen cross sections that have been reduced by corrosion or external forces.

(2) Welding or adding cover-plates to repair structural steel cracks caused by fatigue and vehicle loads.

(3) Retrofitting connections.

c. Rules for adding steel. When steel members are strengthened to carry a specified load, the permissible stresses in the added material must comply with the load design stresses. To properly analyze and design repairs that involve adding

metal, the following rules are applied:

(1) Metal added to stringers, floorbeams, or girders shall be considered effective in carrying its portion of the live loads only.

(2) New metal added to trusses, viaducts, etc., shall be considered effective in carrying its portion of the live load only. The exception is when: The dead load stress is temporarily removed from these members until the new metal is applied, or the dead load stress is applied to the new metal when it is applied.

(3) The added material shall be applied to produce a balanced section, thus eliminating or minimizing the effect of eccentricity on the strengthened member. Where balanced sections cannot be obtained economically, the eccentricity

of the member shall be taken into account in determining the stresses.

(4) Strengthened members shall be investigated for any decrease in strength resulting from temporary removal of rivets, cover plates, or other parts. In some cases falsework or temporary members may be required. Where compression members are being reinforced, lacing bars or tie plates shall be replaced before allowing traffic over the bridge.

11-4. Connections

Primary connections involve the use of welds, bolts, or rivets. However, pin connections and threaded fasteners are used commonly in tension members.

a. Welds. Electric arc welding may be employed subject to the approval of the engineer. In general, welding can be used to repair broken or cracked welds, strengthen rivet connections, and add metal to existing steel members.

(1) *Broken or cracked welds.* Remove all dirt, rust, and paint for a distance of 2 inches from the damaged weld. File or grind down the damaged weld to ensure the weld will bond with the steel surfaces being welded together. For cracks, grind down the cracked weld until the crack is no longer visible then check the weld with dye penetrant to ensure the crack has been completely ground out. Replace the weld. Apply paint or a corrosion protector to welded area.

(2) *Welded rivet connections.* Remove all dirt, rust, and paint for a distance of 2 inches from the steel surfaces in which the weld is to be applied. Apply welds to join the steel surfaces. Where overstressed rivets/bolts can carry the dead load, the weld is designed to carry the impact and live loads. Where overstressed rivets/bolts cannot carry the dead load, the weld is designed to carry the total load. Apply paint or a corrosion protector to welded area.

(3) *Added metal for strength.* Clean the steel surfaces in which metal is to be added and remove any severely damaged or corroded steel portions. Use welds to fill cracks and holes, to replace removed steel portions, or to add coverplates to strengthen individual members. With cracks, ensure the weld penetrates the full depth of the crack. With holes, work the weld from the steel surface to the center of the hole ensuring no voids are allowed in the weld. Replace removed steel portions with steel of the same strength, and, if a torch is used to cut the steel, ensure the edges are ground smooth to ensure a good welding surface. When adding coverplates that will be subjected to compression, ensure the maximum clear spacing

distance (d) is less than 4,000 multiplied by t divided by Fy, where t = plate thickness in inches and Fy = yield stress in pounds per square inch. This prevents local buckling of the attached plate when loaded in compression (figure 11-2). Apply paint or a corrosion protector to welded area and added steel.

b. Rivets. Riveted connections can be repaired using many different methods. The most common repair requirements are for loose or missing rivets or an understrength connection. The repair techniques for each are as follows:

(1) *Loose or missing rivets.* Clean working surface. Replace all missing rivets with high-strength bolts of the same size and draw the nut up tight. This will help support the connection during the repair operations. Tighten or remove and replace loose rivets with high-strength bolts or new rivets. Work on only one rivet at a time to help maintain the proper load distribution to the rivets. Remove and replace high-strength bolts with rivets, one at a time, if the engineer requires rivets for the repair work (high-strength bolts can be substituted for rivets). Check all bolts and/or rivets to ensure tightness. Apply paint or corrosion protector.

(2) *Understrength connection.* Several options exist for an understrength connection:

(a) Rivets or bolts can be added to the existing connection plate.

(b) A longer connection plate can be added to allow for more rivets.

(c) Larger rivets or bolts can be substituted for the existing ones.

(d) Welds can be added to the connection plate of sufficient strength to carry impact and live loads or the total load depending on the condition of the connection.

c. Pin connection. The pin connections discussed herein refer to those used in tension members and military assault bridges. The repair of such con-

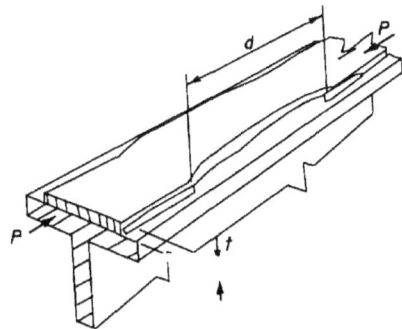

Figure 11-2. Local buckling under compression.

nections may involve the pin itself or the pin housing. To repair or replace the pin or housing, the load must be shifted from the connection through use of jacks or winches. The pin can then be removed from the housing. Repair the housing using welds or by adding metal to build up the connection, as necessary. The pin can either be replaced or a bolt of the same size can be used. The pin connection should be replaced with a welded or riveted connection if it is functioning as a fixed connection or if the bridge is considered a permanent structure.

d. Threaded fastener. Threaded fasteners are used primarily in conjunction with a tension rod or cable. Typical problems with these fasteners involve stripped threads, a rust frozen shaft, or a broken threaded shaft.

(1) *Stripped threads.* Remove the threaded shaft and rethread if necessary. Drill out the threads in the female housing. Place the rethreaded shaft through the female housing and retighten the fastener with a bolt on the backside of the housing.

(2) *Rust frozen shaft.* Clean the shaft and housing of external rust and apply a lubricant to the threads. Attempt to rotate the shaft; if the shaft still refuses to rotate, the fastener must be replaced. Cut the threaded portion of the shaft from the tension member and splice a new threaded shaft to the member. Remove the frozen portion of the shaft from the housing using a torch. Place a bolt on the backside of the housing and pull the tension member tight.

(3) *Broken shaft.* Splice the shaft as shown in figure 11-3.

11-5. Repair of structural members

Structural steel members are generally classified by the function they perform. The primary members are tension members, compression members, beams, and beam-columns. The repair of various structural members is discussed in the following paragraphs.

a. Bars. Structural bars can be either round or rectangular and have a pin or threaded connection. Repair of the bar itself involves tightening the bar to account for elongation, adding metal to the eye of a pin connection, or splicing the bar to repair breakage.

(1) *Tightening adjustable connections or turnbuckles.* Clean the components and attempt to tighten the bar. If adjustments cannot be made, adjustments can be made to the eyes or turnbuckles can be added as follows:

(a) Adjustments to the eyes. For slack of less than 1 inch, the pin connection can be shimmed to take up any slack between the pin and the eye of the bar. Shimming is accomplished by removing the pin from the housing and placing a metal sleeve snugly around the pin and the inside of the eye of the bar. The metal sleeve will take up the slack between the pin and the eye of the bar.

(b) Add turnbuckle. A turnbuckle can be added to a round bar which previously had no adjustable tension system (figure 11-3). This is accomplished by cutting a section out of the existing bar, threading both adjacent ends of the cut, and screwing on a turnbuckle.

(2) *Broken/damaged bars.* Repair broken or damaged bars as follows:

(a) Add turnbuckle. Cut out the damaged area and use the same technique discussed in paragraph 11-5a(1)(b) above.

(b) Splice bars. Broken bars can be spliced by welding bars or plates on both sides of the damaged area as shown in figure 11-3. Bolts and rivets can also be used if desired.

(3) *Reinforcement of eyebars with loop bars.* Eyebars can be reinforced with loop bars in the following manner. Release tension from the eye or consider the added metal for carrying live load and impact loading only. Weld a plate onto the existing eyebar. This plate can be used to attach a supplemental eye (loop or forged bar which reinforces the eye of the bar). Place loop bars around the pin and weld to the steel plate (figure 11-4, part a). Reapply tension and paint repaired area.

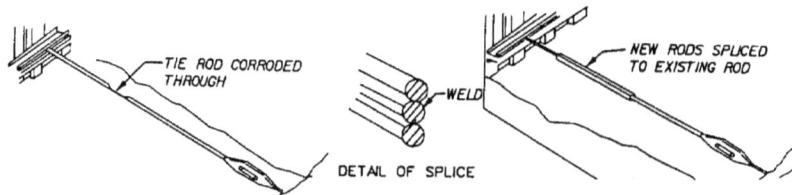

Figure 11-3. Repairing tie rods with splices or turnbuckles.

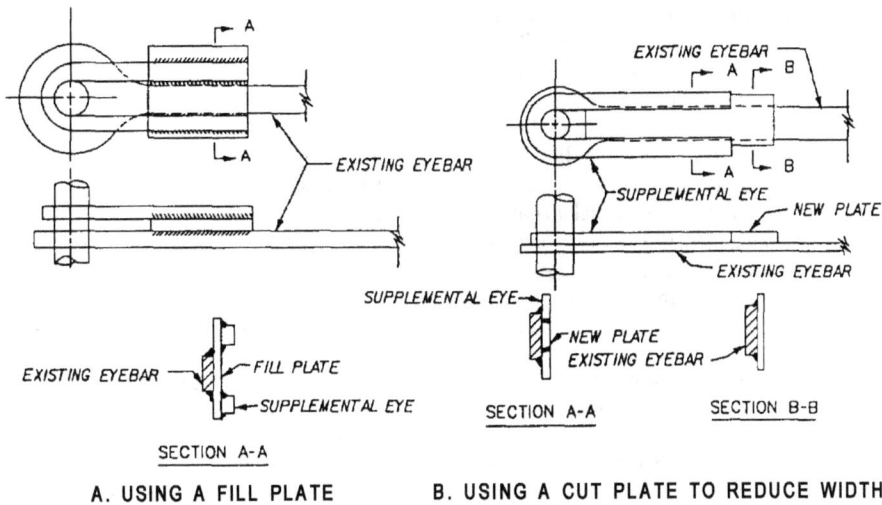

Figure 11-4. Strengthening pin connections using a supplementary eye.

(4) *Reinforcement of eyebars with steel plates (figure 11-4, part b).* Eyebars can be reinforced with steel plates by the following procedure. Release tension from the eye. Place a supplemental eye around the existing eyebar. Slide a new plate between the arms of the supplemental eye. The thickness of the new plate should be equal to that of the supplemental eye. Its width should be equal to the diameter of the pin for the approximate length of the arms of the supplemental eye, and then it should widen to a width greater than the existing eyebar. Its length should be determined by the load carrying capacity requirements of the welds. Weld the supplemental eye, new plate, and eyebar together. Replace the pin, reapply tension, and paint the repair.

b. Hanger plates. Pin and hanger connections are especially vulnerable to corrosion, which may freeze the connection and increase the internal stresses in the hanger, causing cracks to form. If the hanger must be removed for cleaning or repair, a technique which can be used is:

(1) Fashion a temporary hanger to support the dead load normally carried by the hanger. This can be accomplished as follows:

(a) Place a steel plate across the expansion dam on the road surface. The plate should have two holes running the length of the dam and at a distance slightly greater than the flange width of the girder apart.

(b) Run two threaded bars capped with nuts through the holes to the bottom of the girder's bottom flange.

(c) Attach a plate to the bars below the bottom flange.

(d) Tighten the nuts on the top plate to draw the bottom plate up tight to the bottom flange and adjust the tension on the hanger.

(2) Once the temporary hanger is in place, the connection can be cleaned and repaired as follows:

(a) Attach a safety cable to the outside hanger and remove the hanger from the pins.

(b) Clean the pins and hanger thoroughly and repair any damage to these components.

(c) Spot paint the inside face of the hanger with light layers of paint and allow time for the paint to dry while working on the pins.

(d) Lubricate the pins and replace the hanger.

(e) Repaint the outside face of the hanger.

(f) Repeat the procedure for the inside hanger.

(3) Remove the temporary hanger.

c. Cable. The most common damage that occurs to cable is fraying of the steel wires forming the cable. Any repair requiring strengthening the damaged portion of the cable involves removing the tension on the cable. There are three repair techniques that can be used on damaged cables.

(1) If the cable still retains adequate strength to carry its design loading despite the frayed area, then clip off the frayed wires, clean any rust and debris from the cable, and paint or wrap the damaged portion of the cable with a protective coating.

(2) If the cable requires strengthening, then replace the cable, run an additional cable parallel

11-5

to the existing cable to carry the balance of the load, or cut the cable and emplace a turnbuckle.

d. Rolled sections or plate girders. The most common repairs to these sections are repairing cracks in welds and strengthening the sections with cover plates. In some cases, bent steel sections are repaired using a technique called structural steel flame straightening.

(1) *Crack repair.* The typical cracks that occur in rolled sections and plate girders are caused by fatigue and overloading. The cracks generally begin in the flange and propagate through the flange into the web. To properly repair these cracks, the cracks must first be closed and then a weld applied. The importance of closing the crack is to ensure that the dead load is properly transferred across the entire cross section of the steel member. There are various methods which can be used to close these cracks:

(a) Posttensioning the member (see paragraph 11-10).

(b) Jacking the load off the member (see paragraph 12-7).

(c) Using a heated cover plate across the crack and following these steps. Weld a cover plate to the member along only one side of the crack. Heat the cover plate to expand the plate's length. While still heated, weld the opposite side of the plate to the member and allow the plate to cool and contract. This will pull the crack together. Weld the crack and apply a continuous fillet to the cover plate.

(2) *Cover plates.* The cover plate is a relatively easy and inexpensive technique to use in steel repair. It can be used to bridge over a crack and transfer the dead load throughout the cross section of the steel member or to just add strength to the member's cross section. A key dilemma when adding steel cover plates is the method used to introduce dead load stresses into the new material. The most common method used with major members (i.e., W-sections) is to calculate the amount of shorting required to produce the dead-load stresses in the existing member. Holes are drilled into the old and new members in such a way that the new members will be shortened by drifting, as the sections are bolted together. Another method is heating the cover plate after welding one end and then welding the expanded plate into place as discussed. As the plate cools and contracts, stresses will be added to the cover plate. Examples of various uses of the cover plate are:

(a) Extending an existing cover plate to bridge a welded crack (Refer to paragraph 11-4a(3)).

(b) Building up the bottom flange, top flange, or the web of the member to increase the load capacity of a steel member.

(c) Bridging or strengthening steel sections damaged by corrosion or collisions (figure 11-5).

(3) *Rules for using cover plates.*

(a) When one or more cover plates must be renewed, consideration must be made for replacing the defective plates with one plate of adequate size.

(b) Cover plates should be connected by continuous fillet welds, high-strength bolts, or rivets.

(c) Where the exposed surface of old plates are rough or uneven from corrosion and wear, they should be replaced with new plates.

(d) Welded cover plates should be of sufficient thickness to prevent buckling without intermediate fasteners.

(e) Where the web was not originally spliced to resist moment, it may be spliced by adding cover plates or side plates.

(f) Where fatigue cracking can occur at the top of welds on the ends of the cover plate, it is recommended that bolting be used at the cover plate ends.

(g) When cover plates are used in compression members, care must be taken to maintain symmetry of the section to avoid eccentrical loadings.

(4) *Alternative to cover plates.* Where the cost of removal and replacement of the deck would be excessive, as in ballasted-deck bridges, flange sections may be increased by adding full-length longitudinal angles, plates, or channels just below the flange angles. First remove the stiffener angles and then replace them with new stiffeners after the flange steel is added (figure 11-6).

(5) *Stiffeners.* Stiffeners are used to reinforce areas of the beam which are suspectable to web buckling (intermediate transverse stiffeners), concentrated loads greater than allowable stresses (bearing stiffeners), and web buckling due to bending (longitudinal stiffeners).

(a) *Intermediate stiffeners.* Additional stiffeners may be added by riveting, high-strength bolting, or welding angles or metal plates perpendicular to the top flange and web. These stiffeners should not be connected to the tension flange of the member.

(b) *Bearing stiffeners.* These stiffeners can be reinforced by adding angles or plates to the existing stiffeners, grinding the bearing ends of the new parts to make them fit closely, or welding the bearing ends to the flanges.

(c) *Longitudinal stiffeners.* Stiffness may be added to the member by bolting or welding longi-

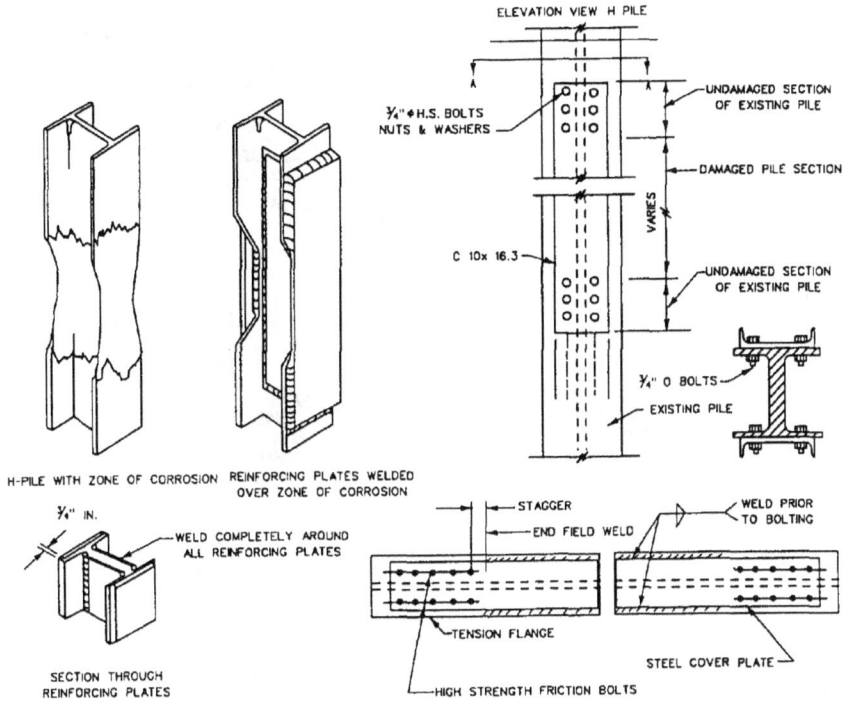

Figure 11-5. Use of cover plates on rolled steel sections.

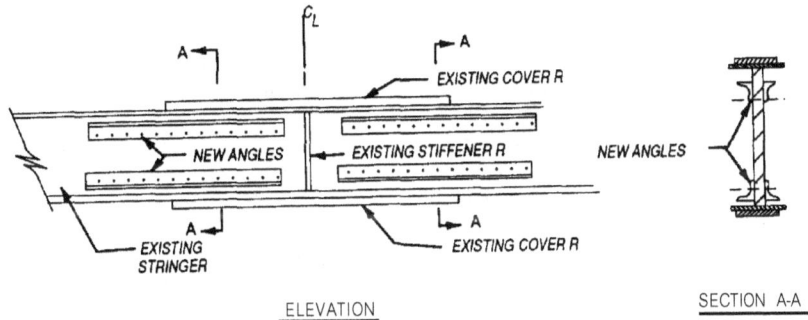

Figure 11-6. Iowa DOT method of adding angles to steel I-Beams.

tudinal angles to the web of the steel member.

(6) *Piles.*

(a) When steel piles require additional support or protection, an integral pile jacket can be placed around the steel piling. The encasement of the steel piles is accomplished by filling a fiberglass form with Portland-cement grout. After the concrete hardens, the fiberglass form remains in place as part of the jacket. The integral jacket provides protection to steel piles above and below the water. If the pile has deteriorated to the point

that additional steel support is required, cover plates can be added to the pile prior to placing the jacket or a reinforced concrete jacket can be designed.

(b) A procedure for installing an integral pile jacket follows (figure 11-7). Sandblast the surfaces clean of oil, grease, dirt, and corrosion (near white metal). Place the pile jacket form around the pile. Ensure standoffs are attached to the form. Seal all joints with an epoxy bonding compound and seal the bottom of the form to the pile. Brace and band the exterior of the form to

11-7

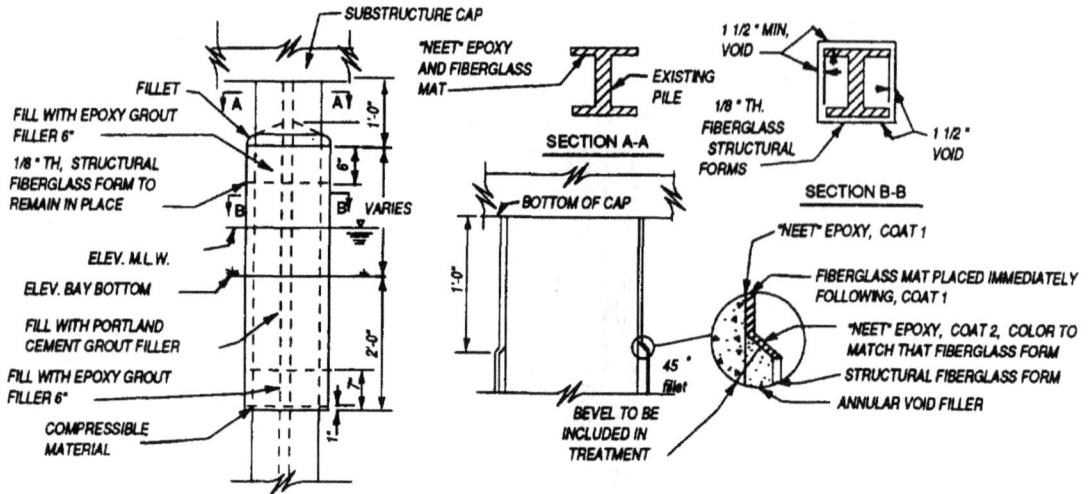

Figure 11-7. Integral pile jacket for steel piles.

hold the form in place. Dewater the form. Fill the bottom 6 inches of the form with epoxy grout filler. Fill the form to within 6 inches of the top with a Portland-cement grout filler. Cap the form with a 6-inch fill of epoxy grout. Slope the cap to allow water to run off. Remove the external bracing and banding and clean off the form of any deposited material.

(7) *Flame straightening.* Flame straightening has been used for over 40 years to straighten bent beams or align members for proper connections. This process involves using "V" heats or triangle heats on flanges in conjunction with jacking devices to shrink the beam in the desired direction. The "V" heat is not heated completely, nor is it gone over again until cooled. The base of the "V" heat should not exceed 6 inches to avoid warping of the flange. The heat used in the process should not exceed 1,200 °F and in most areas will probably be less. It is usually necessary to heat the webs of damaged beams in conjunction with the flange. Overheating and jacking can cause the flange to exceed its yield strength and move vertically, instead of the desired horizontal direction. For these reasons, this process should be conducted with trained and qualified operators only to direct and monitor all heats, jack placements, and to supervise the program's sequence of events.

e. Built-up sections. These sections are primarily used to form truss systems. Cracks due to fatigue and overloading are repaired in the same manner as rolled sections discussed previously. A combination of cover plates, plates, angles, and other rolled sections can be used to reinforce or strengthen the built-up member. Adding metal to these sections increases the cross-sectional steel area in tension members, reduces the slenderness ratio in compression members, and increases the sectional modulus in beams. Examples of strengthened members are shown in table 11-1.

f. Composite sections. Composite action is developed when two load-carrying structural members such as a concrete deck and the supporting steel beams are integrally connected and deflect as a single unit. To ensure that no relative slippage occurs between the slab and beam, the beam's flange is imbedded into the concrete or shear connectors are used to develop a composite action between the surfaces. Three areas may require repair on a composite bridge: the concrete deck, the steel beam, or the composite action between the deck and beam.

(1) *Concrete deck.* Concrete deck damage or faults can decrease the bridge's load capacity by reducing the composite action of the section. The concrete maintenance and repair techniques discussed in chapter 13 should ensure that the deck will carry the compressive forces required by the composite action.

(2) *Steel beam.* Any cracks or damage to the steel section should be repaired (paragraph 11-5d(1)) to ensure that, the correct composite action is received.

Table 11-1. Built-up members

ADDED METAL	DESCRIPTION	FIGURE
COVER PLATE/ PLATES	COVER PLATE IS WELDED TO THE COMPRESSION MEMBER TO IN-CREASE THE STEEL AREA AND REDUCE THE SLENDERNESS RATIO END POST AND BOTTOM CHORDS OF A THROUGH TRUSS ARE BOXED IN WITH STEEL PLATES.	EXISTING MEMBER (CHANNELS & PLATE) · ADDED PLATES · LACING · ADDED COVER PLATES · ADDED SECTION · EXISTING SECTION
ANGLES	ANGLES ARE ADDED TO THE MAIN EXISTING PLATES FOR ADDED STRENGHT.	LACING · ANGLES ADDED TO PLATE
ANGLES & PLATES	PLATES AND ANGLES ARE ADDED TO AN EXISTING MEMBER TO RE-DUCE THE SLENDERNESS RATIO. PLATES AND ANGLES ARE ADDED TO PROVIDE A CENTRAL WEB IN AN EXISTING COMPRESSION MEM-BER.	EXISTING MEMBER (ANGLES & PLATES) · ADDED ANGLES · ADDED PLATES · EXISTING MEMBER (CHANNELS, COVER PLATE, & LACING · NEW MEMBER (ANGLES & WEB PLATE)
ROLLED STEEL	A PLATE/TEE SECTION OR A I, W SECTION IS USED TO INCREASE THE STEEL AREA IN A TENSION MEMBER.	EXISTING DOUBLE ANGLE DIAGONAL · THROUGH BOLT CONNECTION · PLATE (SHOP WELDED) · STRUCTURAL TEE

(3) *Composite action.* Several repair techniques exist to address the loss of composite action:

(a) *Allow for noncomposite action.* Build up the steel beam to carry the load neglecting composite action. Angles can be added to the bottom of the top flange, cover plates can be added to the bottom flange, and/or stiffeners can be used to offset web buckling.

(b) *Repair shear connectors.* The composite action of the section can be reactivated by repairing the shear connectors in the following manner. Use bridge diagrams and plans to determine the location and type of the shear connectors. Remove the concrete from around the shear connector. Repair the existing connectors or replace with welded studs. If welding is not feasible, high-strength bolts can be used as the shear stud as demonstrated in figure 11-8. High-strength bolts may also be added to increase the composite action.

(c) *Deck replacement.* The deck can be completely replaced to renew the bond between the concrete and the shear connectors. A new deck can be poured or a new precast deck can be emplaced with holes cast for the connectors, using grout in the holes around the connectors (paragraph 11-9). Note that the composite action is normally de-

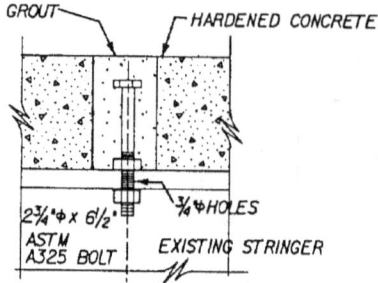

Figure 11-8. Details of double-nutted bolt shear connector.

signed for live and impact loading, unless the dead load is distributed during the concrete curing process.

g. Steel grid decks. Steel decks are often used to increase the live load capacity of bridges when used to replace concrete decks. The primary maintenance and repair problems with steel decks are deterioration from exposure to the weather, weld failure, and skid resistance. Repairs for corrosion and welds are the same for steel decking as with any other structural steel member.

(1) *Concrete fill.* Another method of repair commonly used is to fill or partially fill the steel grid with concrete. The concrete filled grid acts as a reinforced concrete deck. The steel grid provides the steel reinforcement and the concrete fill provides stiffness to help carry the load. This repair technique has several advantages: increased deck strength, reduction in the effect of weathering by limiting the water penetration, support of the welded grid joints, and increased skid resistance. A wearing surface can be added to the deck to provide additional skid and weather resistance. Recommended wearing surfaces are: latex modified concrete (1-inch thickness), asphalt (1.75inch thickness), concrete overfill (1.75inch thickness), or epoxy asphalt (5/8-inch thickness).

(2) *Exodermic deck.*

(a) This deck consists of a thin upper layer (3-inch minimum) of precast reinforced concrete deck panels joined to the steel grating. The panels can be precast and placed on a bridge or added to an existing steel deck using shear connectors welded to the steel grid.

(b) The emplacement procedure follows. Weld shear studs/connectors to the steel grid. Precast concrete deck panels with holes for the shear studs. Place a thin metal, wood, or cardboard plate with a diameter greater than the precast stud hole around the stud. Place the precast concrete deck panels on the steel grid and grout the stud holes. Seal the cracks between the concrete panels.

Section III. MEMBER REPLACEMENT

11-6. Tension members

Any tension in the member must be transferred to a temporary cable support system prior to removing the damaged member (figure 11-9). Attach a cable/turnbuckle system to the supporting frame parallel to the member to be replaced. Transfer the load carried by the steel member to the cable by tightening the turnbuckle. Remove and replace the damaged steel member. Remove the temporary cable support system.

11-7. Compression members/columns

The primary method to support the compressive load during replacement is block and bracing. Emplace a steel or wood section parallel to the member to be removed. Jack the load off the damaged member and place wood or steel shims on the temporary support to transfer the load to the temporary support. In some cases the shims may be driven between the temporary support and the frame to transfer the load. A cable support system wrapped around the temporary support and the frame can be used to hold the member in place during the replacement operation. Remove and

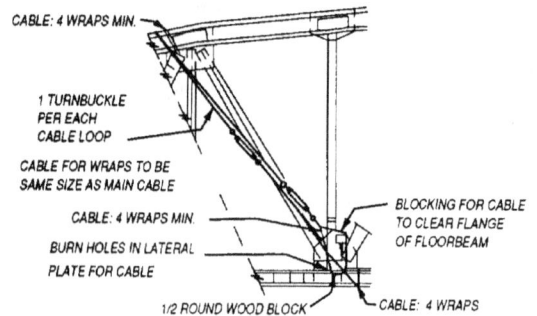

Figure 11-9. Method for relieving stress in tension members.

replace the damaged compression member. Remove the shims and temporary support.

11-8. Beams

The addition or replacement of a stringer is normally performed in conjunction with the replacement of the deck. If the deck is not removed the member must be replaced. from under the bridge. The beam replacement method will depend upon whether the beam/deck system is composite or noncomposite.

a. Noncomposite beams. Place jacks under the deck or under the beams that are not being replaced and jack the deck off the beam to be removed. Remove beam or repair the damaged end of the beam. The beam may be cut to facilitate removal. The beam may be jacked out of position or a lift truck used to lower the beam. Lift or jack the replacement beam into place. To use a lift truck or cable system, a hole must be cut or drilled into the deck to run the cable through. Position the beam on the bearing plates, jack the deck down onto the new beam and check for distress in the beam and deck. Remove jacks, temporary supports, and repair any holes in the deck.

b. Composite beams (figure 11-10). Use jacks to lift the composite beam off its bearing. Burn or cut

the lower portion of the beam (web and bottom flange) from the fixed top flange. Grind the cut area of the top flange smooth. Weld a new stringer to the old stringer top flange for continuity.

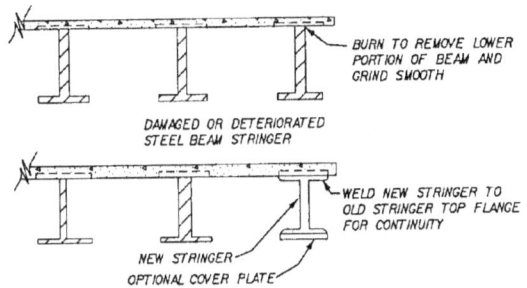

Figure 11-10. Replacement of a steel beam in a composite section.

Section IV. UPGRADE STEEL BRIDGES

11-9. Creation of a composite action

The overall strength of a noncomposite stringer/deck configuration can be greatly increased by making the deck and stringers work together as a single unit using composite action.

a. Concrete deck. The condition of the deck determines how to obtain composite action between the deck and the stringers. If the concrete deck does not need replacing, create a composite action between the deck and stringers by the introduction of shear connectors or studs across the interface of the two materials. If the deck is badly deteriorated, remove the deck and emplace shear connectors or high strength bolts as discussed in section II of this chapter. Another alternative to recasting the deck is to precast concrete panels to be used as bridge decking (figure 11-11). This method reduces the interruption of traffic by allowing the upgrade to be conducted in stages and not having to wait for a recast deck to cure.

b. Steel grid deck. A varying amount of strength can be added to a structure by welding the grid to the top flange of the stringers. The strength increase is based on the stiffness that can be generated by the deck to resist bending. If the grid is filled with concrete, it would act as a concrete deck with the welds providing the interface normally achieved by shear connectors.

11-10. Posttensioning

Posttensioning can be applied to an existing bridge to meet a variety of objectives. It can relieve tension overstresses with respect to service load

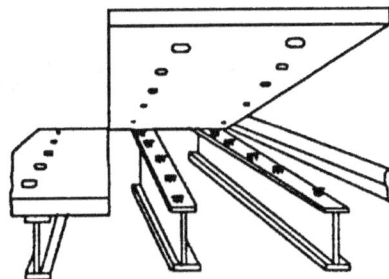

NOTE: SHEAR STUDS SHOWN ARE ACTUALLY ADDED AFTER PRECAST DECK IS POSITIONED.

Figure 11-11. Precast deck with holes.

and fatigue-allowable stresses, reduce or reverse undesirable displacements, and add ultimate strength to an existing bridge. Posttensioning follows established structural analysis and design principles. A design procedure to posttension steel and composite stringer follows:

a. Determine the standards to which the bridge is to be strengthened (loads, allowable stresses, etc.).

b. Determine loads and loads fractions for all stringers (dead load, long-term dead load, live load, impact load).

c. Compute moments at all critical locations for dead load, long-term dead load, and live and impact loading.

d. Compute section properties as required.

e. Compute stress to be relieved by posttensioning at all critical locations.

f. Design posttensioning (tendon force, tendon eccentricity, distribution of axial forces and moments, and tendon length).

$$f = FF (P/A) + MF (Pet/I) \qquad (eq\ 11\text{-}1)$$

where:

f = stress at extreme fiber

P = tendon force

A = area of stringer (composite area if bridge is composite)

e = eccentricity of tendon with respect stringer or bridge

c = distance to extreme fiber

I = stringer moment of inertia (composite section moment of inertia)

FF = force fraction (similar to load fraction for live load)

MF = moment fraction (similar to load fraction for live load)

g. Select tendons, accounting for losses and gains:

$$P_{adjusted} = P / (1 - losses\ \% + gains\ \%)\ (eq\ 11\text{-}2)$$

Assume a 4-percent loss for relaxation of tendon steel, 7-percent loss for potential error in distribution fractions, 2-percent loss for an approximate 10°F temperature differential between tendons and bridge. Assume a 25-percent gain from truck live load.

h. Check stresses at all critical locations.

i. Design anchorages and brackets. Typical ones are shown in figure 11-12. Experience has shown brackets should be about 2 feet long.

j. Check other design factors (beam shear, shear connectors, fatigue, deflection, beam flexural strength, other, as required).

11-11. Truss systems

These systems can be strengthened by adding supplementary members to change to a stiffer system or superposition another system onto the existing structure. In cases where the dead load is increased, the substructure must be checked to ensure that it is adequate for the new loading requirements.

a. Add supplementary members. This technique is most often applied to Warren and Pratt trusses (figure 11-13). The most common use of supplementary members is to reduce the unbraced length of the top chords in the plane of the truss. Additional lateral bracing may be required to reduce the effective length in the plane perpendicular to the truss. The load capacity of the top chords in compression can be increased by 15 to 20 percent using this technique.

b. Doubling a truss.

(1) *Steel arch superposition on a through truss.* A method of upgrading through-truss bridges is to reinforce the existing truss with a steel arch (figure 11-14). A lightweight arch superimposed onto the truss carries part of the dead and live loads normally carried by the truss alone. The truss provides lateral support to the arch. There are two methods of attaching the steel arch to the truss and both methods may require the end panel verticals to be strengthened to ensure adequate lateral support to the arch. In method 1, the existing floorbeams and stringers are assumed adequate to carry the increased live load. In this

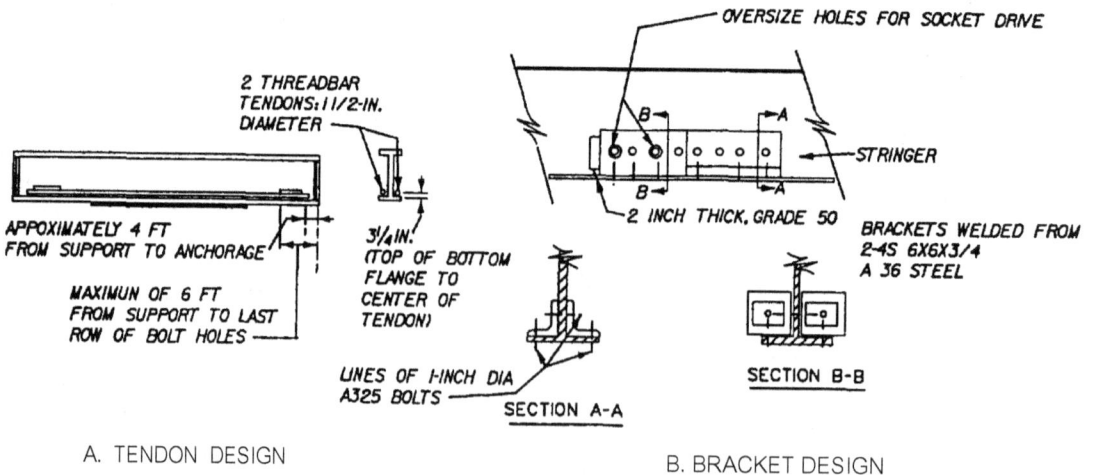

A. TENDON DESIGN

B. BRACKET DESIGN

Figure 11-12. Design of posttensioning.

(A) EXISTING TRUSS

(B) MODIFIED TRUSS

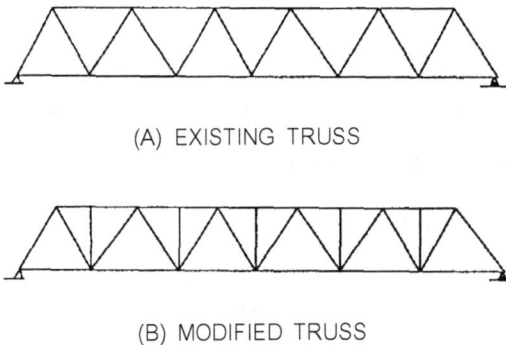

Figure 11-13. Adding supplementary members to a truss frame.

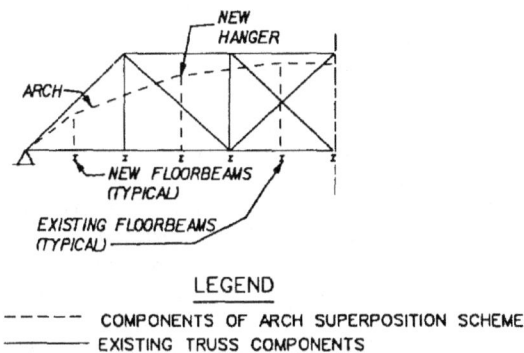

LEGEND

– – – – – COMPONENTS OF ARCH SUPERPOSITION SCHEME
——————— EXISTING TRUSS COMPONENTS

Figure 11-14. Arch superposition scheme.

case, the arch support system is attached to the truss at the existing truss verticals. A conservative design includes the dead load of the arch and the truss, along with the full live load. In method 2, the existing floorbeams and stringers will not carry the increased loading. This requires additional floorbeams at the midpoints of existing floorbeams to increase the capacity of the stringers. The existing stringers can be analyzed as two-span continuous beams, with the new floorbeams providing midspan support. Hangers are attached to the new floorbeams, and the arch is connected to the existing verticals. In the determination of the load-carrying capacity of the reinforced system, the analysis of the arch and truss should be conducted separately, with the two-span continuous stringers providing the loads to each. As an alternative to these methods, the arch can be posttensioned along the bottom chords. Posttensioning will reduce horizontal forces transferred by the arch to the abutments.

(2) *Superimposing a bailey bridge.* Superimposing Bailey trusses onto pony or through-truss bridges is a temporary upgrading method. Normally the Bailey trusses are a few feet longer than

the existing truss and are supported by the abutments. The Bailey trusses are connected by hangers to the existing floorbeams (additional floorbeams may be added and connected to the Bailey truss if the stringers are inadequate). The hanger bolts are tightened until the existing truss shows some movement, indicating complete interaction between the Bailey truss and the existing truss (figure 11-15). The Bailey truss is placed inside the existing truss and blocked and braced against the existing truss to provide lateral bracing to the Bailey truss. This may restrict sidewalk or roadway width. Design data for the Bailey bridge truss are provided in Army TM 5-312. A design procedure for this upgrade is as follows:

(a) Determine additional moment capacity required of the Bailey truss.

(b) Determine initial Bailey truss assembly that meets required moment capacity (Army TM 5-312).

(c) Replace the existing truss with a beam equivalent flexural rigidity.

(d) Replace the selected Bailey truss with a beam of equivalent flexural rigidity.

(e) Consider adding additional floorbeams between existing beams if the stringers are inadequate, and connect the Bailey truss to the old and new floorbeams.

(f) Connect the two trusses rigidly at the points of application of the loads so that they deflect simultaneously an equal amount. The loads carried by each truss can then be determined based on deflected values and flexural rigidity. If the Bailey is inadequate, select another truss configuration with greater moment capacity.

(g) Check required shear capacity.

(h) Design hangers, lateral bracing, and bearing supports. If separate bearing supports are used as shown in figure 11-15, the Bailey truss adds no additional dead load to the structure.

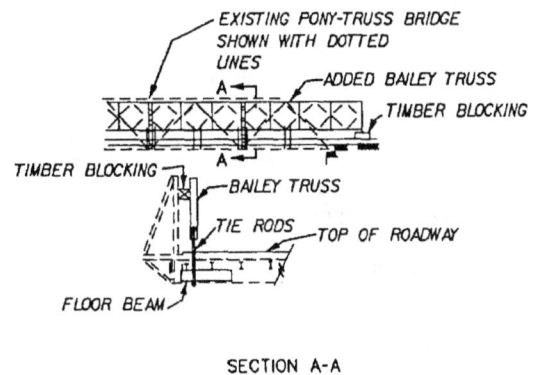

SECTION A-A

Figure 11-15. Reinforcing a pony truss with Bailey trusses.

CHAPTER 12

TIMBER BRIDGE MAINTENANCE, REPAIR, AND UPGRADE

Section I. PREVENTIVE MAINTENANCE

12-1. General

When adequately protected, timber is a very durable building material. The preventive maintenance program for timber involves protection from water and insect damage and the repair of damage from these sources and mechanical abrasion.

a. Water damage. Timbers placed on the ground or at the waterline and continually exposed to wet and dry cycles are subject to rotting. It is helpful to elevate these members with concrete footings or enclose them in concrete. To prevent water from penetrating the wood, the timber itself must be treated as well as any holes or sawed ends of the timber.

(1) *Preservatives for timber treatments.* The penetration of the preservative normally ranges from 1 to 3 inches into the surface.

(a) *Creosote pressure treatment.* This is the most effective method of protection for bridge timbers.

(b) *Pentachlorophenol (penta-oil treatment).* This is a heavy oil solvent applied using pressure methods.

(c) *Inorganic salt solutions.* The salt solution is applied using pressure and provides less water repelling than other treatments. The salt solution can corrode any hardware used to construct the bridge.

(2) *Holes.* Anchor bolts, drift pins, and lag bolts create holes where decay and deterioration often begins. These holes should be protected from moisture penetration by swabbing with hot asphalt or treating with creosote using a pressure bolt-hole technique.

(3) *Timber ends.* The natural opening in the grain of timber ends allows easier water penetration. If the end is cut, it should be painted with a preservative or swabbed with hot asphalt. The ends of the exposed members should be capped with a thin sheet of aluminum, tin, or similar material (figure 12-1).

(4) *Pile ends.* Two options exist for treatment of pile ends.

(a) Drill ¾-inch holes evenly spaced into the pile top 1½ inches deep, fill holes with creosote, and cap with lead sheeting (figure 12-1).

(b) Clamp an iron ring around the top of the pile, pour hot creosote into the ring, and allow the

pile to absorb the creosote, remove the ring, and cap the pile.

(5) *Debris.* Accumulated debris should be removed from any timber surface. Debris holds moisture which will penetrate into the timber member.

(6) *Bark.* The bark of native logs should be removed if it is not removed during construction. This prevents moisture from being trapped between the wood and bark.

b. Insect attack. The insects that attack timber can be classified as either dry land insects (termites, carpenter ants, and powder-post beetles) or marine borers (wood louse or limnoria). The most common treatment for dry land insects is to pressure treat the wood with the proper poison and keep careful watch for reinfestation. Marine borers commonly enter the wood through bruises, breaks, or unplugged bolt holes. The area of infestation generally runs from the mudline to the water level at high tide. The best method of control of marine borers is prevention of infestation. Preventive maintenance for marine borers is conducted for several different stages of infestation.

(1) *Prevent bruises.* Install buttresses to protect the structural members from damage by wave action, current flow, or floating debris.

(2) *Treat damaged wood.* Plug or coat all holes, bruises, and freshly sawed ends with heavy creosote.

(3) *Flexible barrier.* Install a flexible PVC barrier when a pile loses approximately 10 to 15 percent of its cross-sectional area.

(a) *Piles.* Piles should be protected 1 foot below the mudline and 1 foot above high tide (figure 12-2, part a). Sheath the pile with a 30-mil PVC sheet. A half-round wood pole piece is attached to the vertical edge of the PVC sheet to help in the wrapping process. (Note: A pile with creosote bleeding from its surface must first be wrapped with a sheet of polyethylene film prior to installing the PVC wrap to prevent a reaction between the PVC and the creosote.) Staple lengths of polyethylene foam, ½ by 3 inches, about 1 inch from the upper and lower horizontal edges of the sheet. Fit the pole pieces together with one inserted into a pocket attached to the bottom of the other pole. Roll the excess material onto the combined pole pieces and tighten around the pile

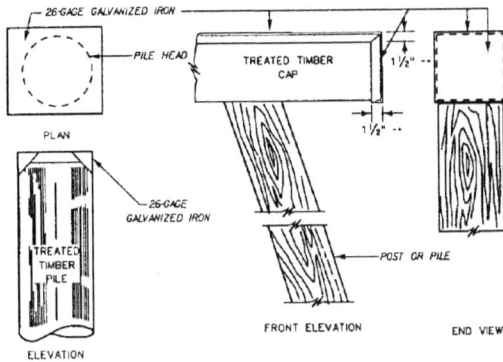

Figure 12-1. Protective covers for timber members.

with a special wrench. Secure the wrap and poles with aluminum alloy rails. Nail rigid plastic bands at the top and bottom directly over where the polyethylene foam is located under the wrap. Install additional bands on equal distance centers between the top and bottom bands.

(b) Braces. Use wrapping when the braces have light damage or to prevent damage from occurring. Remove the bolt which secures the brace to the pile. Wrap the freed end with 20-mil flexible PVC sheeting. Drive the bolt through the wrapping and existing hole. Rebolt the brace to the pile. Wrap the remainder of the brace in the sheeting. Wrap bolt connections as shown in the first steps.

(4) Concrete barrier. When the pile has lost 15

to 50 percent of its cross-sectional area, a concrete barrier can be poured around an existing pile to provide compressive strength and a barrier to marine borers (figure 12-2, part b). To accomplish this, clean the pile to remove foreign materials. Place a tube form (fiberglass, metal, plastic, etc.) around the pile and seal the bottom of the form. Pump concrete grout into the form from the bottom to the top, thus forcing the water out of the form as the grout is placed.

c. Mechanical abrasion and wear.

(1) This is an important consideration when timber is used as piles or decking material. Piles are subject to abrasion and wear from the current and tidal flow and the effect of debris carried by these flows. Methods available to help prevent problems in this area are:

(a) Emplace buttresses on the upstream side of the piles.

(b) Encase the pile in concrete or steel.

(2) Decks are susceptible to wear and abrasion at locations where sand and gravel are tracked onto the bridge. Two methods to protect the wood deck are:

(a) Treadways. Sheet plates or wood planks are placed in the wheel lines to provide a wearing surface.

(b) Bituminous surface treatment. A liquid asphalt is applied to the deck and covered with gravel. This protects the deck from wear; it also seals between the planks and contributes to water-tightness. Note that the asphalt wearing surface will not bond to freshly creosoted timber. It is recommended that the deck should be surfaced only after a period of at least 1 year. This permits

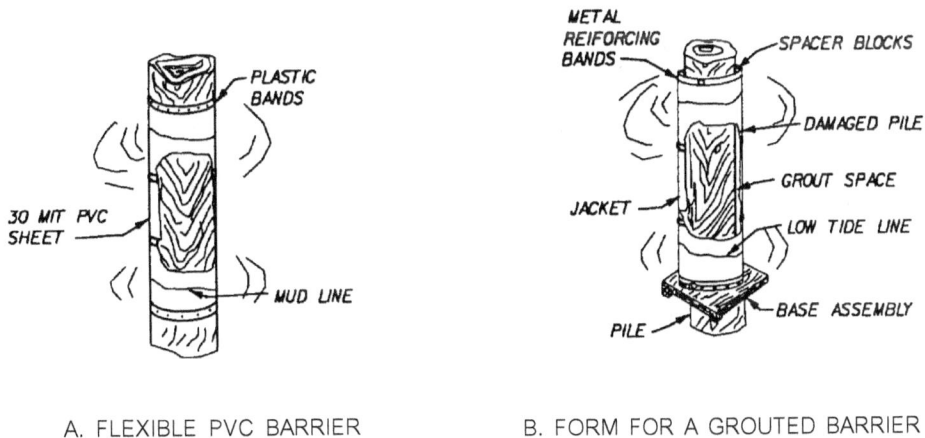

A. FLEXIBLE PVC BARRIER   B. FORM FOR A GROUTED BARRIER

Figure 12-2. Flexible PVC barrier installed on a timber pile.

the creosote to leach from the surface before the asphalt is applied.

12-2. Fire protection

To prevent tire, covered water barrels with a bucket to distribute water can be placed in convenient locations. The danger of fire can also be reduced by a thin coating of tar or asphalt covered with sand, gravel, or stone chips.

Section II. REPAIR AND STRENGTHEN TIMBER MEMBERS

12-3. General

Rcpair methods for wood and timber structures are generally directed at correcting one or more of the following problem areas: fungi and/or insect attack, deterioration, abrasion, and overload. The most common repairs for timber structures are retrofitting timber connections, removing the damaged portion of the timber member and splicing in a new timber, and removing the entire member and replacing it with a new member. Common rules for timber repair are:

a. There should be at least 1/8-inch clearances between timbers to allow the timber to dry properly.

b. When native logs are used for construction, all bark should be removed to reduce moisture penetration into the logs.

c. Green or wet timber shrinks considerably when seasoned. Repeated wetting and drying also causes dimension changes as great as 5 to 10 percent in the direction perpendicular to the growth rings. Frequent renailing and tightening of bolts is necessary.

d. Care should be taken when using new and salvaged wood together to carry loads because of a difference in the sag and shrinkage of the members. The repair should avoid using new and old stringers in the same panel.

e. Wood shims or wedges should be made from heart cypress, redwood, douglas fir, or of the same material as the member.

f. Replacement members must have the same dimensions as the existing member accounting for shrinkage.

g. Always treat drill holes and cut ends, to prevent water or insect damage.

12-4. Connections

The typical timber connection relies on some type of hardware to connect timber members. However, with the improvements made in glues and lamination, glues are becoming a viable option that can be used to enhance the connection. They provide a bond between the wood surfaces and prevent moisture penetration into the connection.

a. Bolts, drift pins, and screws. The two most common repairs required for these connections are the replacement of the bolts due to rust or damage and retrofitting the bolt hole in the timber. After removal and inspection of the existing bolt or screw, proceed as follows:

(1) No deterioration (provides a snug fit for the bolt). Inject a wood preservative into the hole and replace the bolt.

(2) Slight deterioration (loose fit for the bolt). Drill out the hole to a size that provides a good wood surface. (Note: The diameter of the hole should not be increased to the extent that the wood cross section will no longer carry the design load.) Inject a wood preservative. Replace the bolt or screw with a larger size to tit the increased hole diameter.

(3) Moderate deterioration (hole provides no bearing on the bolt and boring out the hole would reduce the connections load capacity). Remove the deteriorated wood from around the edge of the hole. Inject a wood preservative, and, if possible, coat with tar or creosote. Attach steel plates with holes corresponding with those in the timber connection across the connection to bridge the damaged area. Place bolts through holes in the plates and tighten.

(4) Heavy deterioration (connection is beyond repair). Cut off the damaged portion of the member and splice on a new portion, or replace the members which form the connection.

b. Wood scabbing. Scabbing is used to join members together or splice repaired timber members into existing components. Exterior plywood can be used for light loads. However, in most cases standard sawed timbers are used. The first step in any repair of this nature is to check the scabbing for soundness and replace if required. If the scabbing is in good shape, it can be tightened in various ways:

(1) Remove loose nails or screws and replace with larger size.

(2) Add nails or screws to the existing scabbing.

(3) Drill through the scabbing and member and emplace bolts.

c. Steel connector plates. Connection plates are made of light-gauge galvanized steel plates in which teeth or plugs have been punched. If the plate must be replaced, the wood surface may be too damaged for the teeth of the new plate to provide a shear interface. In this case, wood scabbing can be used to replace the steel plate or a new steel plate can be emplaced using nails to provide the interface. A loose plate can also be tightened and strengthened by adding nails or screws to the plate.

d. Nails, spikes, and screws. These are the most common hardware items used to form wood connections. When these items become loose the only options are to:

(1) Replace the nail or screw with a larger one.

(2) Drive or screw the connector back in place, and add nails or screws to help carry the load of the loose connector.

e. Deck connectors. Timber decking is connected to steel stringers with floor clips (figure 12-3) or nails driven into the under side of the decking and bent around the top flange of the steel member (figure 12-3). Composite action is achieved between a concrete deck and timber stringers by either castelled dapping, consisting of ½ to ¾-inch cuts in the top of the stringer; castelled dapping in conjunction with nails or spikes partially driven into the top of the stringer; lag bolts at a 45-degree inclination to horizontal; or epoxies.

12-5. Repair of graded lumber

When lumber is damaged to the point that the structural integrity of the member is in question, scabbing or slicing of the member may be required to bridge the damaged area.

a. Scabbing. Scabbing can be used when the member is moderately damaged and the addition of the scabbing allows the member to carry its design load. The method of scabbing used depends upon the type of timber member:

(1) *All members.* Clean and treat damaged area. Relieve the member of any load, if possible. Place scabbing on the sides and/or the top and bottom of the member to transmit the load across the damaged area. The scabbing can be of a like wood, steel plate with bolt holes, or a steel plate connector.

(2) *Beams and stringers.* A steel plate can be scabbed onto these members to repair longitudinal cracks as follows (refer to figure 12-4): drill holes through the deck on either side of the stringer at a minimum distance equal to the beam depth past the damaged area; drill an additional set of deck holes past the other end of the damaged area; place draw-up bolts through the holes that extend past the bottom of the existing beam (use washers or support straps to help prevent pull through); place a steel retaining plate on the bottom of the stringer and hold it in place with support straps and nuts on the draw-up bolts; and tighten the retainer plate in place. This process will strengthen the damaged area by providing a steel cover plate across the damage, closing any slits or cracks in the wood, and by creating a composite action between the stringer and the deck.

(3) *Piles.* A reinforced concrete scab or jacket can be placed around a partially deteriorated pile to restore the strength. The procedure is the same as placing protective cover around a pile with the

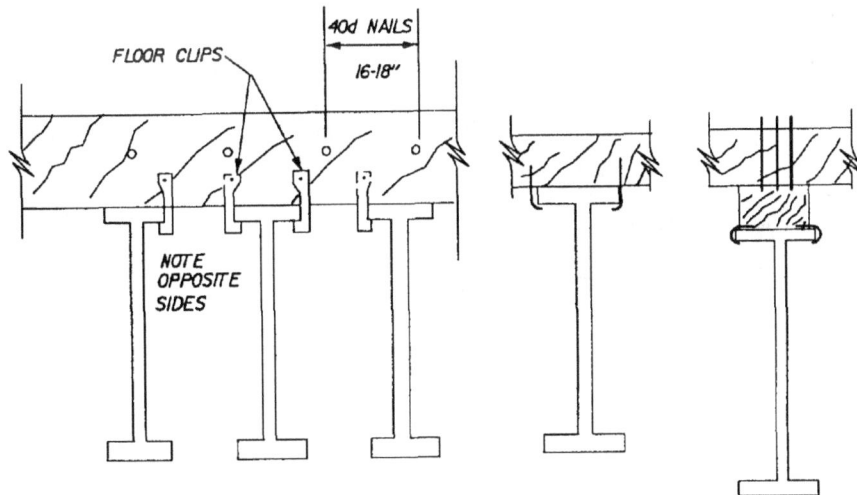

Figure 12-3. Common deck connectors.

exception that reinforcing bars are placed inside the form for added strength (see paragraph 12-1).

(4) *Caps.* Caps can be scabbed to extend the bearing area where the bottoms of stringers or laminated decking has deteriorated over the cap. The repair consists of attaching 6-inch-thick timbers with a depth equal to the cap to each side of the existing cap as follows (figure 12-5). Use a template to lay out 4-inch O-ring connectors on the cap and scabs. Cut O-rings and drill ¾-foot holes through the cap and scabs. Insert O-rings in the cap. Position scabs and clamp or bolt into place. Insert bolts through scabs and cap using O.G. washers on both ends of the bolt, and tighten into place.

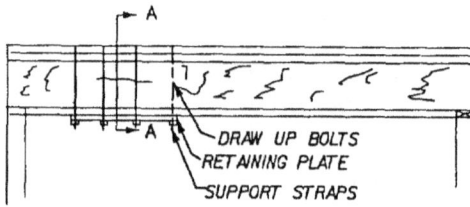

b. Splicing. Splicing is required when the lumber is too severely damaged to carry its design load (figure 12-6). To splice, remove the damaged portion of the wood member. Treat the freshly cut end of the member and the replacement member. Then, nail, screw, or bolt scabbing onto the existing timber and replacement to join them together.

12-6. Repair of piles

When piles are damaged or deteriorated to the point that the structural integrity of the pile is in question, it may be more advantageous to repair the existing pile than to drive a replacement. The key to pile repair is that the existing pile must have good bearing as part of the foundation. The

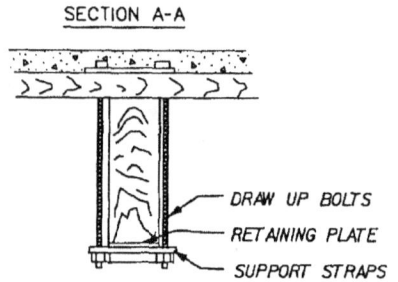

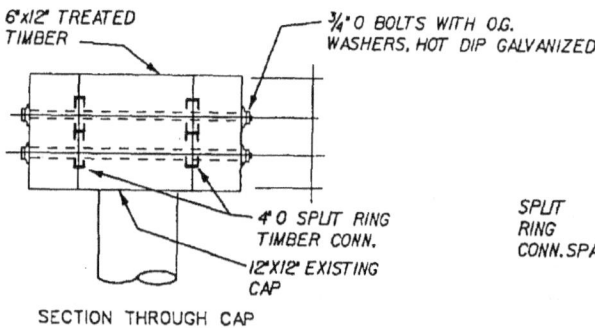

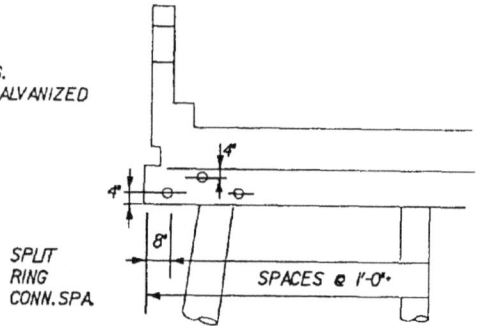

Figure 12-4. Repair of cracked or split stringers

Figure 12-5. Timber cap scabs provide additional bearing.

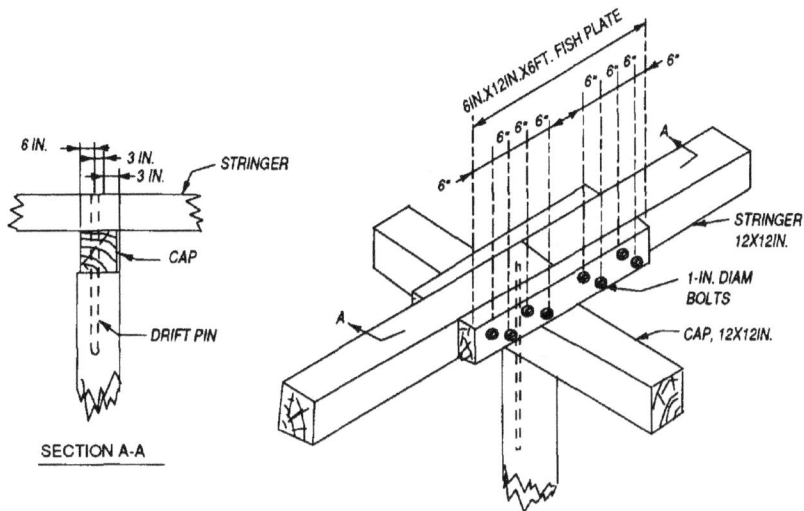

Figure 12-6. Stringer splice.

method of repair chosen depends upon the type of pile damage:

a. *Pile damage extending below the waterline.*

(1) *Single-pile steel reinforced splice (figure 12-7, part a).* Sever the pile approximately 2 feet below the mudline. Cut a stub pile the length of the defective portion. Connect the ends of the stub pile and the existing pile with a center drift pin or dowel with a ¾-inch diameter and 18 inches long. Reinforce the joint by placing three angle sections 2 feet long (1.5 inches by 1.5 inches or 2 inches by 2 inches) at each third point around the circumference of the pile connection. The angles will be held in place by a minimum of four lag screws (4 inches long) per angle.

(2) *Multiple pile steel reinforced splice (figure 12-7, part b).* Sever the piles about 2 feet below the mudline. Place a mudsill on the portion of the piles left in place. Secure even bearing on the piles. Tamp earth between piles for an even sill support. Ensure that the cross section of mudsill equals the bent cap and extends 3 feet beyond the piles. Connect pile stubs between the mudsill and cap using the drift pins and angles described for the single pile. If the pile damage extends above the waterline, sever the pile about 2 feet below the damage and proceed as previously described for a below-water single pile. Seal the joint with creosote or asphalt.

b. *Single pile damage above the groundline (figure 12-8).* Remove the soil around the pile to below moisture line. Construct cribbing or place struts for support jacks. Place jacks and jack up

cap ½ inch to 1 inch. Cut the dowel connecting the cap to the pile. Cut the deteriorated pile off below the permanent moisture line. Make cut at a right angle to the centerline of the pile. Cut the new pile ¼ inch longer than the removed section. Place the concrete form and reinforcing bars into position. The form must allow a minimum of 6 inches of cover around the pile. The form can be made from steel culvert or steel drums and need not be removed. Place the new pile section on the existing pile stump. Ensure an even contact at the bearing points and check to make sure the rebar has not shifted. Pour Class II concrete into the form and slope the top of the pour to allow water to run off. Reattach the cap to the pile with a dowel or an exterior steel plate.

c. *Settlement and bearing loss of a pile due to deterioration.* Place cribbing or struts adjacent to the pier and jack the stringers off the cap to an elevation ½ inch higher than desired. Cut the tops off the decayed pile. Cut a shim ¼ inch less than the space between the cap and the pile head and treat the pile head and shim. Place the shim into position. Lower jacks and fix the shim into position (figure 12-9). Toenail the shim to the pile. Dowel through the cap and repair. Nail fish plates across the repair and remove the cribbing.

d. *Damage to fender piles.* Piles that have been broken between the top and bottom wales (figure 12-10) can be repaired as follows: cut off the pile below the break; install a new section, secured with epoxy; fit a strongback into position behind

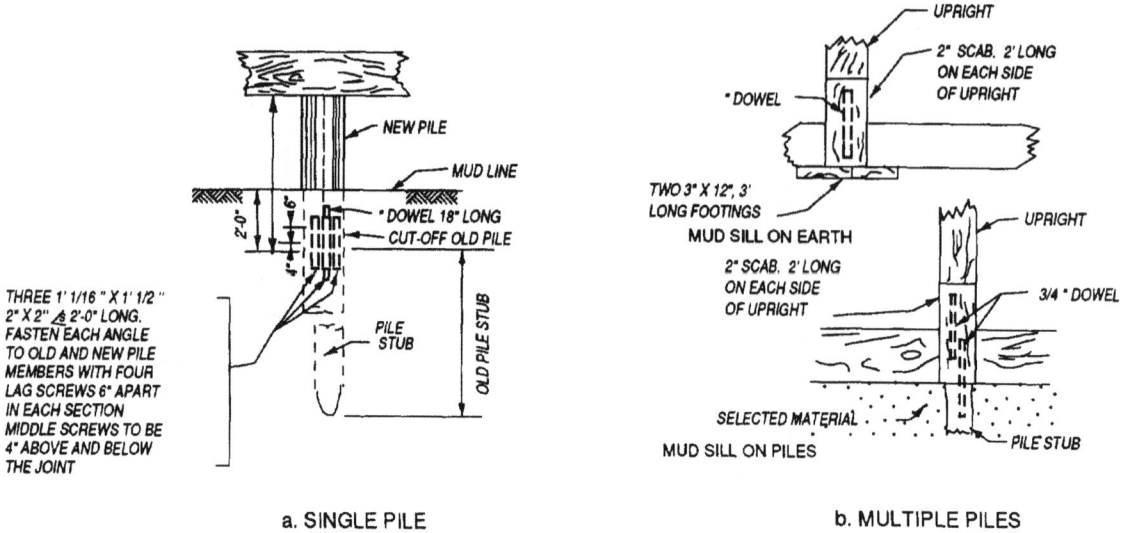

a. SINGLE PILE

b. MULTIPLE PILES

Figure 12-7. Timber pile repair.

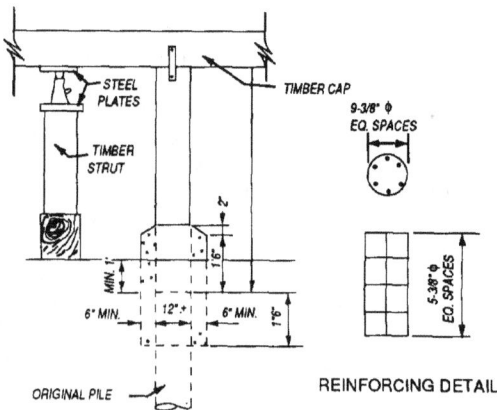

Figure 12-8. Concrete jacket supporting a timber pile splice.

the pile with bolts; and connect a metal wearing shoe on the front side of the pile.

12-7. Repair of posts

Damaged bent posts can be repaired in much the same way as single piles or through splicing across the damaged section.

12-8. Repair of sway bracing

Repairs involving sway bracing may include the erection of bracing to help stabilize a pier or the replacement of existing bracing. When bracing is desired, measurements should be taken and the timber cut and treated prior to the repair operation.

a. Installation of bracing. Temporarily attach bracing to piling in its final position using galvanized nails. Locate and drill bolt holes through the bracing and pile. The holes should be the same diameter as the bolts. Treat all holes with a hot oil preservative. Install bolts, washers, and nuts and tighten into place.

b. Repair sway bracing. Locate the end of the deteriorated or damaged brace. Cut off the damaged portion to the nearest pile. Take the required measurements for the new bracing and cut the timber to size. Drill bolt holes in the new bracing, rebolt the bracing to the piling using existing holes when possible, and drill new holes in the piles to realign bracing. Treat all timber cuts and holes with hot oil preservatives followed by a coating of hot tar.

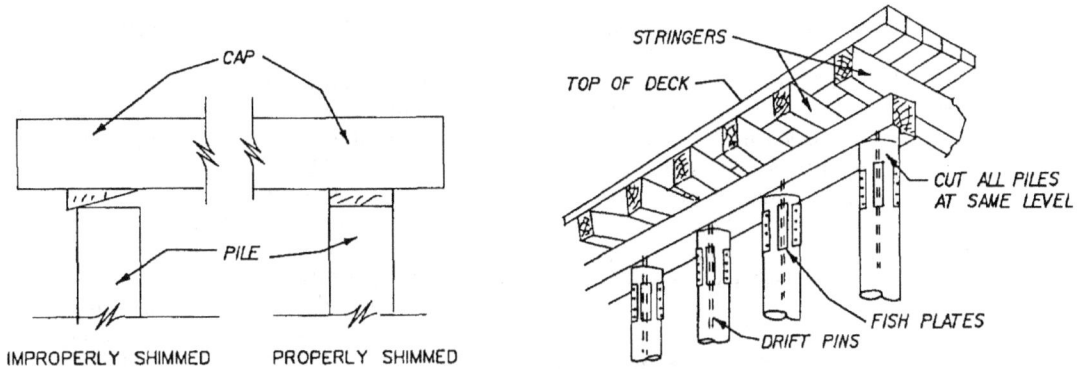

Figure 12-9. Shimming timber piles.

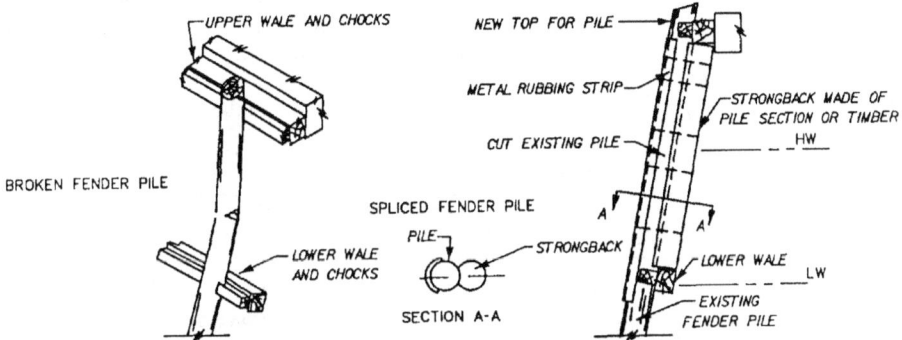

Figure 12-10. Fender pile repair.

Section III. MEMBER REPLACEMENT

12-9. Replacement of tension timber components

These members can be replaced using the same procedures outlined in paragraph 11-6 for steel members.

12-10. Replacement of compression timber components

a. *Piles.* There are three methods that can be used to replace a damaged pile:

(1) Pull the existing pile and drive a new pile.

(2) Drive a new pile along side the existing pile as follows (figure 12-11, part a): locate the centerline of the stringers nearest to the pile to be replaced, cut a hole through the deck to drive the new pile, remove bracing which interferes with pile driving. Set pile at a slight batter so that it is plumb when pulled into position. Drive to a specified bearing. Install U clamps and blocking around the pile being replaced or on adjacent piles if the existing pile has no load carrying capacity. Place a

jack on the block system, and jack the cap off the piles. Cut the new pile ¼ inch below the cap, and place copper sheeting on the pile head. Pull the pile into position. Lower jacks and dowel or strap the cap to the pile. Repair and replace bracing and deck.

(3) Leave the existing pile in place and add new piles as follows (figure 12-12): cut holes in the deck to drive piles, and drive piles on either side of the damaged pile, perpendicular to the bent. The new pile should off set at least one pile width to one side of the damaged pile. Cut the tops of the new piles and place a support across the tops of the new piles to form a cap under the existing pile cap. The two caps should be in contact with each other. Use wedges to ensure that the load is transferred to the new bent.

b. *Bent post.* Replace as follows: install a temporary support parallel to the damaged post, remove the existing post, install a new post and wedge into place to ensure proper load transfer, drive a drift pin through the pile cap, wedge into the post to secure into position or nail the wedge into place, cut the wedges off flush with the cap, post and connect the post to the sill, and cap with scabs.

c. *Caps.* To replace caps, temporary or false bents should be constructed to support the deck and floor during the replacement operation (figure 12-11). The new caps should be given a free, even bearing on each pile or post support. The cap should be fastened to each post or pile by drift pins and spikes.

12-11. Replacement of flexural timber components (stringers)

Stringers should be lined up to a true plane for new or replacement wood (figure 12-13). Stringers in good condition and of proper size may be salvaged and reused. New and salvaged stringers should not be used together in any one panel because of differences in sag and shrinkage be-

tween the new and old stringers. When selecting material for new stringers, it should be of the same width and depth as the other stringers in the panel; however, in case of emergency, the best available size may be used temporarily. The stringers should be wedged as tightly against the deck at the center of the span as they are at the end; do not attempt to fit to the deck sag by adzing. As the new stringers acquire sag, the wedges will be tightened to compensate. The procedure for adding or replacing stringers is as follows:

a. *Under-the-deck method (figure 12-13).* Place two jacks on each cap in adjoining bays next to the stringer being replaced. Use steel plates between the jack faces and the timber. Stop traffic during the jacking operation and jack the deck up ¼ to ½ inch to clear the stringer. Cut a wedge out of one end of the stringer, and bevel the corners on the other end. If the replacement timber is warped, place the camber up to provide bearing on all the deck members. Place the stringer's wedged end on

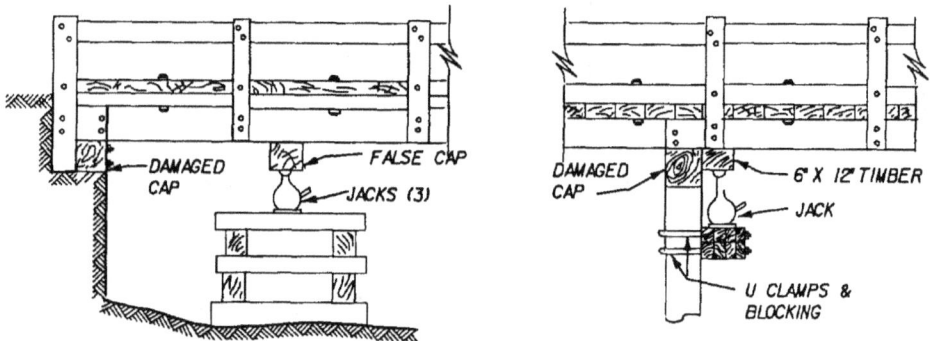

A. JACKING FROM CRIBBING.

B. JACKING FROM A BENT PILE.

Figure 12-11. Jacking methods for timber cap replacement.

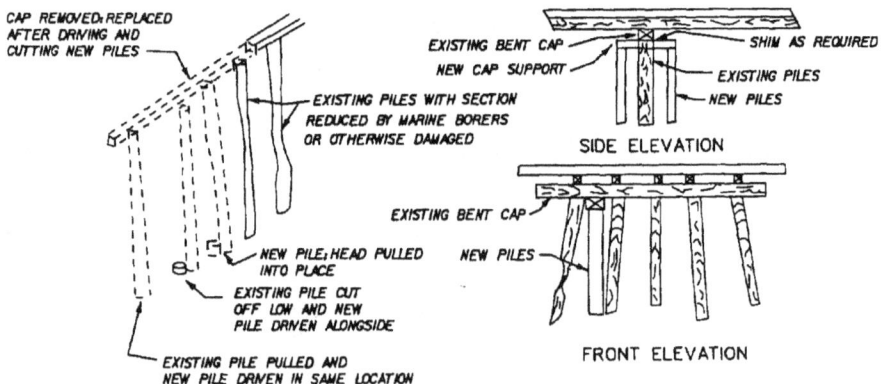

Figure 12-12. Pile replacement methods.

its cap, and push it far enough onto the cap to allow the opposite end to be placed on its cap. Lift the beveled end of the stringer onto the cap. Anchor a "come-along" to the cap under the beveled end, and attach the cable to the wedged end of the replacement stringer. With the come-along, pull the new stringer into a final position that provides an equal bearing on each cap and wedge the under side of the stringer to provide contact with the deck. Remove the support equipment, and nail the wedge into position in such a fashion that the nails can be removed and the wedges adjusted to account for sag.

b. Above-the-deck method. Cut the deck along each side of the stringer to be replaced. Remove enough decking to allow one end of the stringer to rotate on the cap enough for the other end to clear the far side cap when it is lifted. Lift the far end of the stringer with a cable through a precut hole in the deck clear of the cap. Jack or pull the stringer into final position. Replace the decking.

12-12. Replacement of timber decking

a. Timber flooring. Floor planks should be laid with the heart side down because it is more resistant to decay. A ¼-inch spacing should be provided between planks for drainage, expansion, and air circulation. Structural grade hardwood planks, 6 to 10 inches wide, are preferred as decking material because wider planks have a tendency to curl. All nails or spikes should be driven so that their heads are imbedded into the plank. Planks of the same thickness should be placed adjacent to each other with a full, even bearing on the stringers. Wedges should not be used to level flooring because they are easily dislodged and leave the flooring in a loose and uneven condition.

b. Wheel guards. Replacements should use the same size sections as the original and be fastened with the same bolt spacing. The bolts should be extended through the riser or scupper blocks and floor planks (figure 12-14).

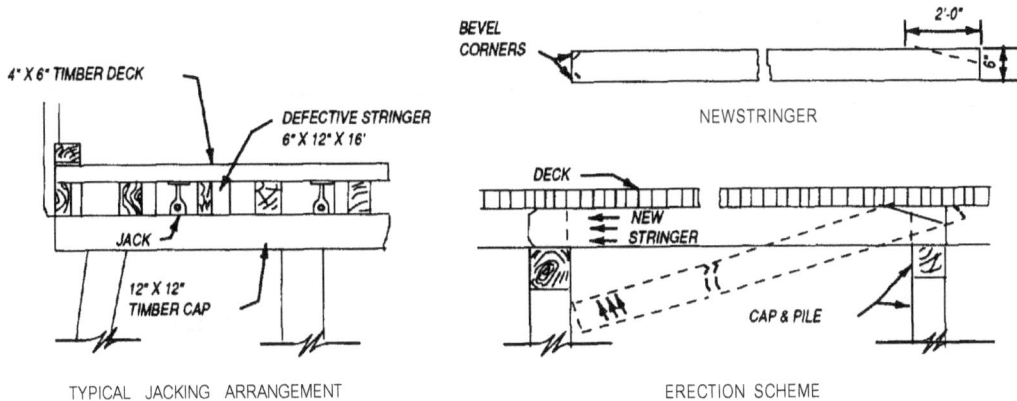

Figure 12-13. Belowdeck timber stringer replacement.

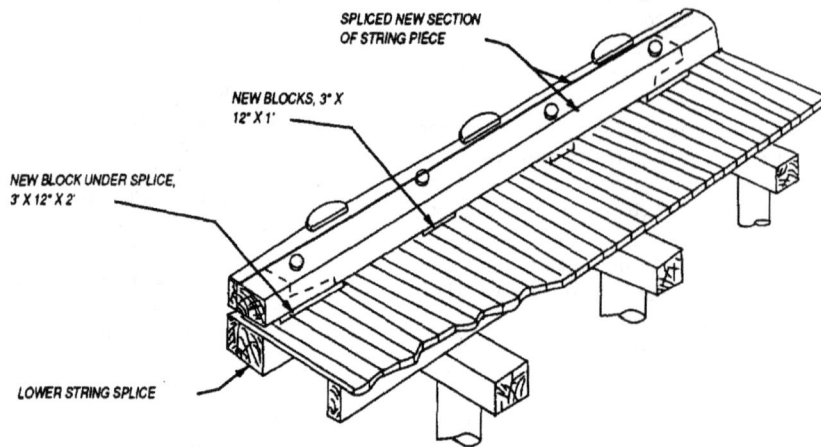

NEW BLOCK UNDER SPLICE, 3' X 12' X 2'

NEW BLOCKS, 3" X 12" X 1'

SPLICED NEW SECTION OF STRING PIECE

LOWER STRING SPLICE

Figure 12-14. Splicing in a wheel curb.

Section IV. TIMBER BRIDGE UPGRADE

12-13. Strengthen intermediate supports (piers)

a. Bent post. Bent posts can be strengthened by nailing 2-inch thick planks to the cap and sill and the post between them. The planks should be approximately the width of the pier component to which it is nailed. This has the effect of increasing the cross sectional wood area carrying the compressive loads of the bridge.

b. Concrete encased piles. Paragraph 12-4 discusses the process of encasing a timber pile in concrete to shore up a timber splice. The same type of technique can be used to extend the reinforced concrete casing the full length of the timber pile. The procedure involves the same steps required in jacketing steel or concrete piles (Refer to paragraphs 11-4 and 13-5).

c. Helper bents.

(1) This system can be used to strengthen an entire pier section. It is used when existing piling have lost their bearing and settlement occurs, when the load capacity of the piling is in question, or when the load bearing capacity of an existing pier is increasing. Timber piles are driven through the deck parallel and adjacent to the pile bent and topped with a cap to interface with the stringers. In essence, a new pier is constructed adjacent to the existing pier to help carry the loads (figure 12-15).

(2) Helper bents are installed in the following manner. Locate the centerline of stringers or

beams and mark the positions of the helper bent piles parallel to the pier. Cut holes in only one traffic lane at a time. For timber decks, make a cut along the centerline of the stringers and remove enough decking to drive the pile. For reinforced concrete decks, remove sufficient concrete in a square pattern to drive the piles. Cut the reinforcing steel at the center of the hole and bend out of the way. Set the piling and drive to the required bearing. Cut the pile ¼ inch above the existing cap (make allowances for grade differentials due to settlement). Place cover plates over the deck holes and move to the next position. Repeat the same operation until all the piles are driven. Jack up the superstructure ½ inch using the existing pier. Place a timber cap onto the pilings. Lower the superstructure onto the caps of the new piers and strap the caps to the pilings. Shim between the superstructure and the cap as required to obtain the proper bearing. Remove the deck plates and repair the deck. Construct cross-bracing on the new pile bents and between the bents for intermediate bents.

12-14. Shorten span length

The strength increase obtained from shortening a timber span is fairly large due to the short spans generally associated with timber bridges. Shortening is accomplished by adding one of the following:

a. Intermediate piers. Paragraph 10-22b.

b. A-Frame. Paragraph 10-22c.

c. Knee brace. Paragraph 10-22d.

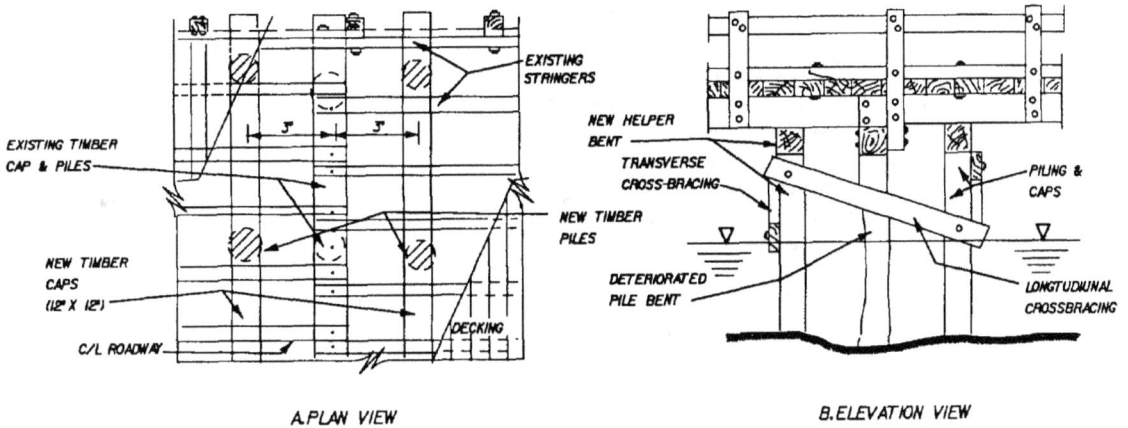

Figure 12-15. Diagrams of an intermediate helper bent.

12-15. Posttensioning

Many types of posttensioning can be applied to timber structures. The most common and effective methods are the king post, which counteracts bending, and the external stirrup, which reduces shear effects.

a. King post.

(1) In effect, the king post shortens the effective length of the reinforced span. One technique that can be used to install the king post to a timber bridge is as follows (figure 12-16):

(2) Install a ¾-inch threaded steel eyebolt through the stringer at an equal distance from each end of the member. Steel plates with a hole for the eyebolt should be used under the eyebolt nuts to help prevent pull through. Note that the eyebolts should be positioned in a low moment area of the stringer and the internal moments generated by posttensioning should be checked. Run a steel cable through the eye of the eyebolt and use u-bolts to tie the cable to the eyebolt. Attach a king post to the underside of the beam at midspan. The king post can be of steel or wood and should have an eye to run a steel cable through. Run a cable through the eye of the king post, and connect the ends of the cable to the cables from the eyebolts using a turnbuckle. Tighten the turnbuckle to provide posttensioning to the stringer.

b. External stirrups. External stirrups can be applied to timber stringers to provide shear strengthening in much the same way as stirrups are used to reinforce concrete. Small channels or angles are used as tie plates across the top and bottom of the stringer, and threaded bars tightened into holes in the tie plates form the external reinforcement (paragraph 13-7).

12-16. Add stringers

Additional stringers can be installed in an existing bridge to redistribute the bridge loading. The same technique used to replace stringers can be used to install additional stringers under an existing deck (paragraph 12-10).

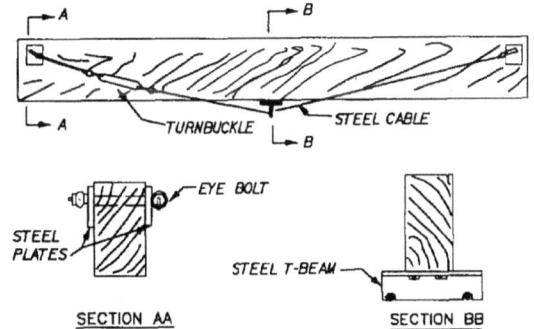

Figure 12-16. Timber beam strengthened using king post posttensioning.

12-17. Strengthen individual members

Timber members can be strengthened by the addition of steel cover plates to counteract flexure and shear as follows (figure 12-17): Apply a field preservative treatment to the timber and paint the steel cover plates. Attach the flexural cover plate to the timber member using lag screws. Attach shear cover plates to both sides of the member using through bolts. Note that proper spacing of lag screws or bolts will ensure that composite action between the plate and timber stringer occurs.

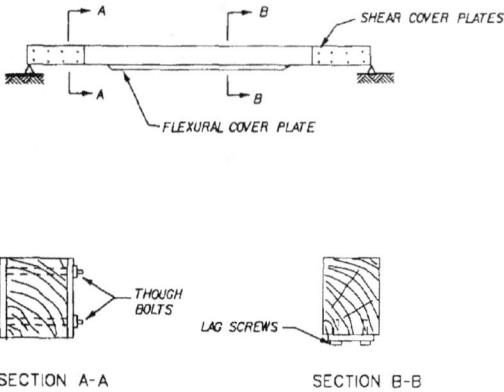

Figure 12-17. Steel cover plates used to reinforce a timber beam.

CHAPTER 13

CONCRETE BRIDGE MAINTENANCE, REPAIR, AND UPGRADE

Section I. PREVENTIVE MAINTENANCE

13-1. General

Preventing concrete deterioration is much easier and more economical than repairing deteriorated concrete. Preventing concrete deterioration begins in the design of the structure with the selection of the proper materials, mixture proportions, concrete placement, and curing procedures. Even a well designed concrete will generally require follow-up maintenance action. The primary types of maintenance for concrete are surface protection, joint restoration, and cathodic protection of the reinforcing bars. Surface maintenance involves the application of coatings for protective purposes. Joint problems are usually treated with one of a variety of types of joint sealers, and cathodic protection involves the use of anodes connected to the reinforcing bars which will deteriorate in place of the reinforcing bar.

13-2. Surface coating

a. General. Surface coatings are applied to concrete for protection against chemical attack by alkalies, salt solutions, or other chemicals. The actual need for a coating must first be established, and then the cause and extent of any deterioration, rate of attack, and environmental factors must be considered for selecting the right coating for the job. A variety of coatings and sealants are available for waterproofing and protecting concrete surfaces. Among the products are several types of oil and rubber resins, petroleum products, silicones, and other inorganic and organic materials. Some of these products have been successful in protecting new concrete from contamination by deicing salts and other harmful environmental agents. They have generally been unsuccessful at stopping the progression of already contaminated concrete.

b. Surface water repellents. This type of coating helps prevent or minimize scaling from the use of deicers. This is a low-cost treatment that provides a degree of protection for non-entrained concrete or is added insurance for air-entrained concrete placed in the fall and subject to deicing salts during the first winter.

(1) Linseed oil. A mixture of 50 percent linseed oil and 50 percent mineral spirits is normally used. The mixture is applied in two applications on a dry, clean concrete surface. The surface coating should be less than 5 mils, and a test strip should be used to help determine the application rate. The normal application rate is 40 square yards per gallon for the first application and 65 square yards per gallon for the second application. This treatment should last for 1 to 3 years.

(2) Silicone. Silicone has been used on concrete to minimize water penetration. Care must be taken where moisture has access to the backside of the wall and carries dissolved salts to the front face where it is trapped by the silicone. Silicone oxidizes rapidly and is somewhat water soluble. Treatments are required every 1 to 5 years.

c. Plastic and elastomeric coatings. These coatings form a strong, continuous film over the concrete surface. To be effective in protecting concrete, the coating must have certain basic properties: the adhesive bond strength of the coating to the concrete must be at least equal to the tensile strength of the surface concrete; the abrasion resistance must prevent the coating from being removed; chemical reactions must not cause swelling, dissolving, cracking, or embrittlement of the material; the coating should prevent the penetration of chemicals that will destroy the adhesion between the coating and concrete; for proper adhesion, the concrete must be free of loose dirt particles, oil, chemicals that prevent adhesion, surface water, and water vapor diffusing out of the concrete.

(1) Epoxies. Epoxies are used, with a solid content from 17 to 100 percent as a clear sealant, with coal or tar mixed as a mortar. As with most thin coatings and sealants, a protective overlay or cover is required if they are exposed to traffic wear or abrasive forces.

(2) Asphalt. Asphalt is used as a protective overlay for bridge decks. This surface provides water protection and a protective wearing surface.

13-3. Joint maintenance

Little maintenance is required for buried sealants because they are not exposed to weathering. Most field-molded sealants require some type of periodic maintenance if an effective seal is to be maintained. Minor touchups in field-molded sealants can usually be made with the same sealant. Where the failure is extensive, it is necessary to remove and replace the sealant. The sealant can be removed using handtools or on large projects by routing or plowing using suitable tools. Sawing can be used to enlarge the joint to improve the shape factor for the new sealant. After the joint has been cleaned it can be resealed. For more information on joints see paragraph 12-12.

13-4. Cathodic protection

The only remedial procedure other than replacement that has proved effective in stopping the corrosion process in contaminated concrete is cathodic protection. Compared with the cost of replacement, it is normally much less expensive. There are two types of cathodic protection systems that can be used to prevent corrosion in reinforcing bar. The systems are the impressed current system and the galvanic system.

a. Impressed current system. This system uses low-voltage, high-amperage direct current from an external power source to make the reinforcing bar into a cathode. This system is most effective when installed during the construction of decks, piers,

stringers, and abutments. During the installation ensure that the reinforcing bars are in good contact with one another; this will allow the current to flow through the reinforcing system. This system can be used to protect the exposed reinforcing bar; however, it is usually not an economical option in this special case. A common power source used in the system is a rectifier which provides DC power from an AC source.

b. Galvanic system. This system requires no external power source but uses anodes of a special alloy to generate the current required to suppress the corrosion process. Three common alloys are zinc, magnesium, and aluminum. A procedure used to protect exposed reinforcing bars in concrete piles follows (figure 13-l): Clean an area on the exposed reinforcing bar on which to attach a zinc anode (figure 13-1, part a). Attach one 7-pound anode to every 6 feet of reinforcing bar exposed to brackish or salt water (figure 13-1, part b). Note that additional site testing may be required to adjust the weight of the anode to ensure a 2-year life for the anodes. Less brackish water or fast moving water will affect this life expectancy.

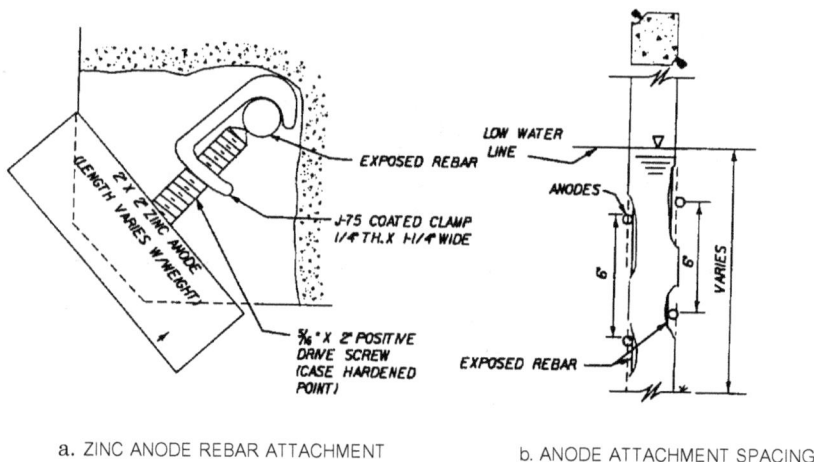

a. ZINC ANODE REBAR ATTACHMENT

b. ANODE ATTACHMENT SPACING

Figure 13-1. Cathodic protection for reinforced concrete piles.

Section II. REPAIR AND STRENGTHEN

13-5. General

The repair and strengthen methods discussed in the following paragraphs apply only to conventionally reinforced concrete and specifically do not apply to prestressed concrete. Refer to paragraph 13-17 for a discussion of prestressed concrete members. In concrete repair it is imperative that

the causes and not the symptoms of the problem are dealt with. The types of concrete damage, the causes, and the means of determining the causes are discussed in chapters 5 and 8. The repair of deteriorated concrete can be categorized into repairs suited for cracking and those suited for spalling and disintegration. Repair of deteriorated concrete is required when: (1) The deterioration affects the structural design performance of the bridge; (2) The deterioration exposes the reinforcing steel to corrosive action; (3) The lack of a repair will make the bridge unsafe for traffic under normal operating conditions. (Examples: potholes, extensive spalling of the deck, damaged parapets, etc.)

13-6. Crack repairs

The large variety of crack types prevents a single repair method. Active cracking may require strengthening of the concrete across the crack to prevent further crack expansion and the application of a flexible sealant that will expand with the crack. Dormant cracks basically require bonding across the crack for the load carrying portion of the concrete and sealing in all other areas.

a. Conventional reinforcement. This method is primarily used to bridge isolated cracks in the load-bearing portion of the structure for active and dormant cracks. This repair bonds the cracked surfaces together into one monolithic form (figure 13-2). Proceed as follows: clean and seal the existing crack with an elastic sealant applied to a thickness of 1/16 to 3/32 inch and extending at least ¾ inch on either side of the crack, drill ¾-inch holes at 90 degrees to the crack plane, fill the hole and crack plane with epoxy pumped under low pressure (50 to 80 psi), and place a reinforcing bar (No. 4 or No. 5) into the drilled hole with at least an 18-inch development length on each side of the crack.

b. Prestressing steel. This technique uses prestressing strands or rods to compress the crack to close it. (Refer to paragraph 13-12c).

c. Drilling and plugging. This repair consists of drilling down the length of the crack and grouting the hole to form a key that resists transverse movement of the section. This technique is most effective on isolated cracks that run in a straight line and are accessible at one end (figure 13-3). To accomplish, proceed as follows: drill a hole (2- to 3-inch diameter) centered on and following the crack its full depth, and clean out the drill hole. The specifics of the remaining procedure are given for:

(1) *Dormant cracks.* Fill the drill hole and crack with a cement grout. If watertightness is required over the bond transfer, fill the hole with asphalt or a like material. When both watertightness and a keying action is required, drill two holes along the crack and fill one with grout and the other with asphalt.

(2) *Active cracks.* Fill the hole with precast concrete or mortar plugs set in bitumen. The bitumen is used to break the bond between the plug and the hole to prevent cracking of the plug by subsequent movement of the crack.

d. Dry packing. This process consists of ramming or tamping a low water content mortar into a confined space. This technique is effective in patching holes with a high depth-to-area ratio or dormant cracks. To accomplish, proceed as follows: Undercut the area to be repaired so that the base width is slightly greater than the surface width. For dormant cracks, a slot should be cut along the surface of the crack 1 inch wide and 1 inch deep with a slight undercut. A sawtooth bit is a good tool to use for this purpose. Clean and dry the slot. Apply a cement slurry bond coat of equal parts of cement and fine sand to the faces of the slot. Place dry pack mortar in the slot in 3/8-inch layers and compact each layer with a hard wood stick working from the middle out or a T-shaped rammer for larger areas. If the mortar becomes spongy during the compaction process, allow the surface of the dry pack to stiffen and then continue the compaction. Fill the slot completely and strike flush with

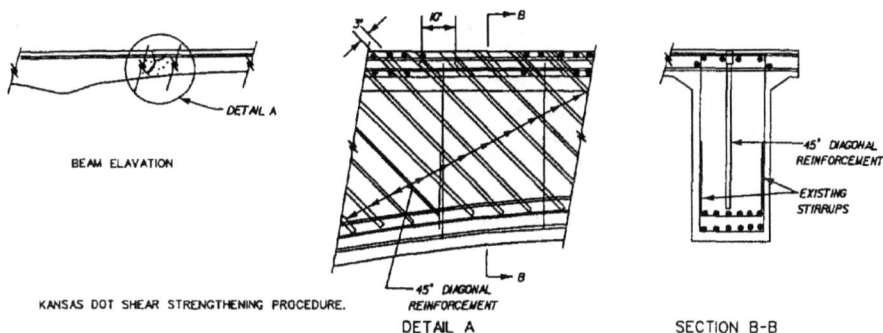

BEAM ELEVATION

KANSAS DOT SHEAR STRENGTHENING PROCEDURE.

DETAIL A

SECTION B-B

Figure 13-2. Reinforcing bars inserted 90 degrees to the crack plane.

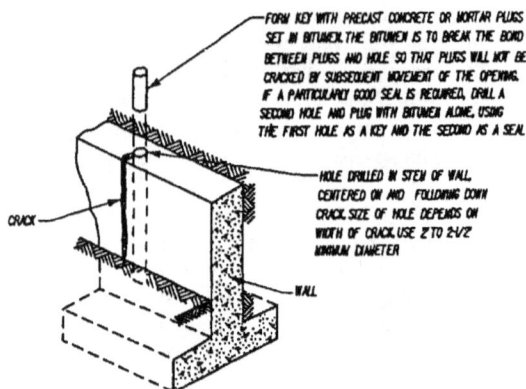

Figure 13-3. Crack repair by drilling and plugging.

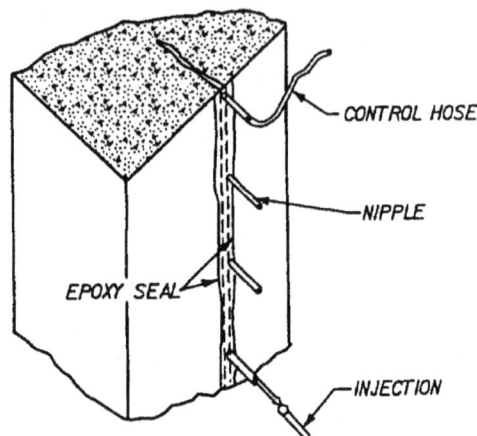

Figure 13-4. Epoxy injection used to seal cracks.

the concrete surface using a board or the hardwood compaction stick.

e. Epoxy injection.

(1) Cracks as narrow as 0.002 inch can be bonded by injection of epoxy along the crack (figure 13-4). This technique also has been used to repair delamination in bridge decks.

(2) To accomplish, proceed with the following steps. Clean the crack of all oil, grease, dirt, or substances that may retard the bonding process. Seal the surface of the crack by brushing an epoxy along the surface of the crack and allowing it to harden. If high-pressure injection is required, cut a V-shaped groove along the crack ½ inch deep and ¾ inch wide, and fill the groove with epoxy. Install epoxy injection ports. There are three procedures in current use for the installation of injection ports:

(a) Drilled holes with fittings. This method is used in conjunction with V-grooves and involves drilling ¾-inch-diameter holes to a depth of ½ inch below the apex of the groove. An injection fitting (small-diameter pipe, plastic tubing, valve stem, etc.) is bonded into the hole.

(b) Bonded flush fittings. When cracks are not V-grooved, injection ports are bonded flush with the concrete face. This type of port would be seated directly into the crack, and the face of the port would be flush with the concrete face.

(c) Interruption in seal. A portion of the seal is omitted from the crack. For this method, the epoxy injector must have a gasket system that covers the unsealed portion of the crack and allows the epoxy to be injected into the gap without leaking.

(3) After installation of the injection ports, mix the epoxy to conform to the current American Society of Testing and Materials specification for Type I, low-viscosity grade epoxy. Inject the epoxy

into the crack using hydraulic pumps, paint pressure pots, or air-actuated caulking guns. Vertical cracks should be injected starting with the port at the lowest elevation and working up. Horizontal cracks can proceed from either end of the crack and run to the far end of the crack. Remove the epoxy seal by grinding or some other means and paint over the injection ports with an epoxy patching compound.

f. Flexible sealing. This repair method involves routing and cleaning the crack and fitting it with a field-molded flexible sealant. It is used for active cracks in which the crack is the indication of a joint requirement in the concrete, and the formation of a joint does not impair the capacity of the structure. To install, proceed as follows (figure 13-5): Rout out the active crack to the dimensions that comply with the width and shape factor requirements of a joint having the same movement. Clean the crack and routed area by sandblasting and/or water jetting. Apply a bond breaker at the bottom of the slot to allow the sealant to change shape without forming a stress concentration on the bottom of the sealant. Common bond breakers are polyethylene strips and pressure sensitive tape. Fill the slot with a suitable field-molded flexible sealant in accordance with the proper American Concrete Institute (ACI) specification. Narrow cracks can be sealed without routing by applying a bond breaker over the crack and overlapping the sealant across the bond breaker to seal the crack (figure 13-5).

g. Routing and sealing. This method is used to seal dormant cracks that do not affect the structural integrity of the bridge member. The seal prevents water from reaching the reinforcing steel. Proceed as follows (figure 13-6): Rout along the

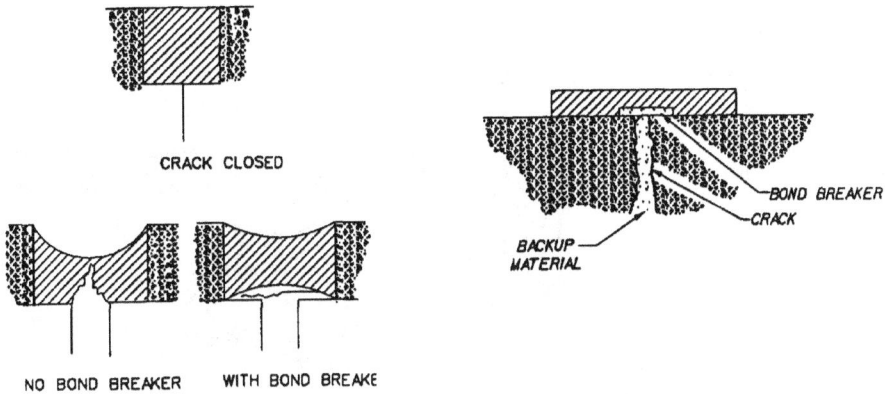

Figure 13-5. Flexible seals used in concrete crack repair

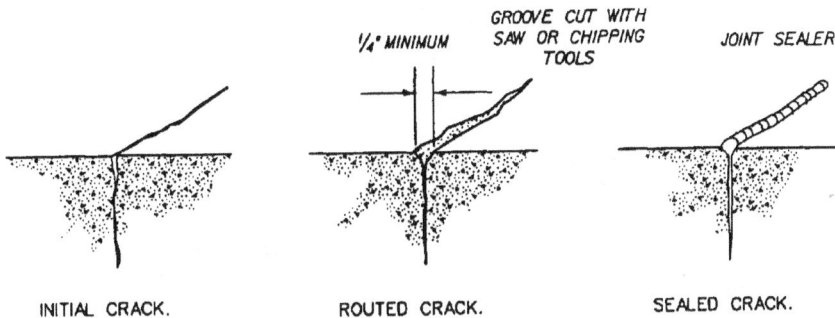

Figure 13-6. Conventional procedure for sealing dormant cracks.

crack to provide a minimum surface width of ¼ inch. Clean the cut and allow the surface of the cut to dry. Apply sealant (ACI 504R).

h. Grouting (hydraulic-cement). Dormant cracks can be repaired with portland cement containing slag or pozzolans for strength gain. The grout can be sanded or unsanded. Proceed as follows: Clean the concrete along the crack. Install built-up seats (grout nipples) at intervals astride the crack to provide contact with the pressure injection apparatus. Seal the crack between the grout nipples with a cement paint or grout. Pump grout into the crack through the nipples. Maintain the pressure for several minutes to ensure good penetration of the grout. The grout should have a water-cement ratio of one part cement to one to five parts water. The water-cement ratio can be varied to improve the penetration into the crack. Chemical grouts can also be used. They consist of solutions of two or more chemicals that react to form a gel or solid precipitate as opposed to cement grouts that consist of suspensions of solid particles in a fluid. Guidance regarding the use of chemical grouts can be found in EM 1110-2-3504.

i. Stitching.

(1) Stitching is the process of drilling holes on both sides of the crack and grouting in stitching dogs (U-shaped steel bars with short legs) that bridge the crack (figure 13-7). Stitching is used to reestablish tensile strength across the crack. Adjacent sections to the cracked section may require strengthening to prevent a crack from forming in the adjacent sections of the concrete.

(2) Install stitching as follows: Drill a hole at each end of the crack to blunt it and relieve the stress concentrations. Clean and seal the crack, use a flexible seal for active cracks. Drill holes on both sides of the crack. The holes should not be in a single plane and the spacing should be reduced near the ends of the crack. Clean the holes and anchor the legs of the dogs in the holes with a nonshrink grout or an epoxy. The stitching dogs should vary in length and orientation to prevent transmitting the tensile forces to a single plane.

(3) The following considerations should be made when using stitching: Stitch both sides of the concrete section where possible to prevent bending or prying of the stitching dogs. Bending members may only require stitching on the tension side of the member. Members in axial tension

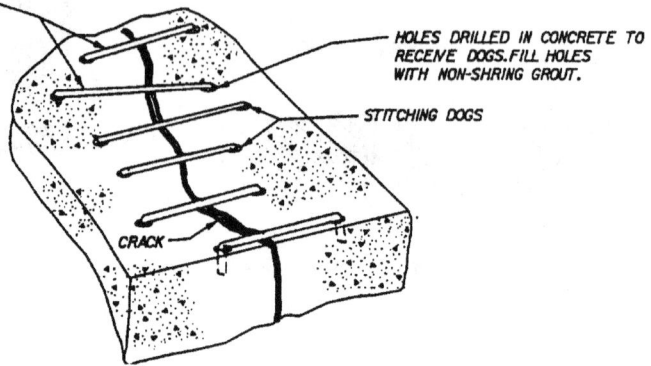

Figure 13-7. Reinforcement of a crack using stitching.

must have the stitching placed symmetrically. Stitching does not close the crack, but can prevent its propagation. Stitching that may be placed in compression must be stiffened and/or strengthened to carry this force, such as encasement of the stitching dogs in a concrete overlay.

13-7. Spall repair

Spall is repaired primarily by removing the deteriorated concrete and replacing it with new concrete of similar characteristics. The process involves the following:

a. Analyze the structure to determine the effect of removing the deteriorated concrete down to sound concrete will have on the structure. Determine the need for and the design of any shoring and bracing required to support the structure during the repair.

b. The concrete must be removed down to sound concrete or to a depth where the patch is at least ¾ inch thick. Sharp edges, at least 1 inch deep, should be formed around the area to be patched to avoid feather-edging the concrete patch. It is also advantageous to make the bottom of the removed concrete areas slightly wider than the surface to form a keying effect with the new concrete patch. If a large surface area is to be overlaid with new concrete, a minimum of ¼ inch should be removed from the surface. The edges of the overlay should be chipped or cut at about 45 degrees to prevent entrapping air under the overlay. Tools that can be used to remove concrete are: jackhammers, diamond saws, rotary head cutters, high-pressure water jets, thermal lances, and hydraulic splitters. Clean sound surfaces are required for any repair operation, and the absolute minimum amount of

concrete to be removed is all unsound concrete, including all delaminated areas.

c. The patch area must be cleaned to remove all debris from the concrete removal process. The existing concrete surface and reinforcing steel should then be blast cleaned. The repair is cleaned again and inspected. Any aggregate particles that have been cracked or fractured by scarifying or chipping should be removed to sound concrete.

d. Patches should be reinforced with wire mesh attached either to reinforcing bars or dowels to secure the patch to the old concrete. Loose reinforcing bars should be tied at each intersection point to prevent relative movement of the bars and repaired concrete due to the action of traffic in adjacent lanes during the curing period. If new reinforcement is required, an adequate length to attain a lap splice (30 times the bar diameter) must extend from the existing section. If a proper splice is not possible, holes must be drilled into the existing concrete and dowels or anchors installed.

e. An interface must be established between the existing and new concrete. Options for this include:

(1) *Epoxy bonding.* Ensure the surface is clean, dry, and free of oil. Apply the epoxy agent to the prepared surface.

(2) *Grout or slurry.* Clean the prepared surface and saturate with water. Remove all freestanding water with a blast of compressed air, and apply a thin coat of grout.

f. After surface preparation, the new concrete must be promptly applied to the repair and finished.

g. The new concrete should be moist cured for a minimum of 7 days to prevent drying shrinkage and to allow proper strength development. This is most easily accomplished by covering the patch with plastic or wet burlap.

h. Shotcrete can also be used to replace concrete in spalled areas. Shotcrete is a mixture of portland cement, sand, and water shot in place by compressed air. It is best used for thin repair sections (less than 6 inches deep) or large irregular surfaces. Shotcrete requires a proper surface treatment similar to that required for a concrete overlay and no form work is needed to confine the mix. This makes shotcrete useful in the repair of vertical walls, beams, and the underside of decks. This technique requires specialized training and guidance on the use of shotcrete as a repair material is given in EM 1110-2-2005.

13-8. Joint repair

The maintenance and minor repair of joints is covered in chapter 10 of this manual. Deterioration around a joint in concrete may require a repair of the concrete around the joint in conjunction with resetting the joint. Some of the specialized joint repairs are as follows:

a. Joint sealant repair. This repair can be used when the sealant has failed, but the premolded joint filler is still in good condition. The repair procedure is as follows: remove the existing sealant (with tools such as a mechanical joint cleaner or joint plow); score the joint walls with a pavement saw; clean the joint with a mechanical brush and remove debris; place sealant in accordance with the manufacturer's installation procedure.

b. Expansion joint seal. This seal can be used to seal open joints, replace failed preformed joint material, and reseal joints sealed with elastomeric seals. The installation procedure is as follows. Determine the required elastomeric seal size in accordance with the guidelines in table 13-1:

Table 13-1. Elastomeric seal size guidelines

Existing Joint Width (in.)	Width of Saw Cut (in.)	Size of Joint Seal (in.)
½	1	1⅝
1	1½	2¼
2	2½	3½
2½	3	4
3	3½	5

Clean the joint. A high pressure water jet is useful in removing any debris or existing joint material in the joint. Saw cut the joint across the bridge to provide a uniform dimension in the repair joint (figure 13-8, part a). Blow out the joint with compressed air to remove any standing water, dirt, gravel, etc. Brush an adhesive/lubricant onto the inner joint faces. Position the seal over the joint (figure 13-8, part b). Compress the bottom portion of the seal and press into the joint. Ensure the seal is seated to its proper depth and alignment. To make an upward turn with the seal, proceed as

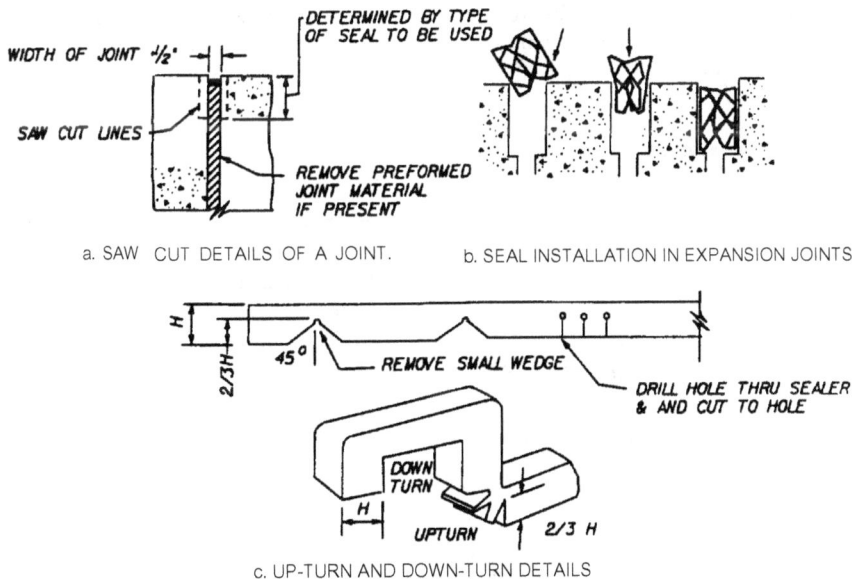

a. SAW CUT DETAILS OF A JOINT.   b. SEAL INSTALLATION IN EXPANSION JOINTS

c. UP-TURN AND DOWN-TURN DETAILS

Figure 13-8. Expansion joint repair with elastomeric seals.

follows (figure 13-8, part c). Drill three ½-inch-diameter holes on line in the elastomeric seal. The holes should be spaced at intervals of 1/3H (H = height of the seal) on lines 2/3H from the bottom of the seal. Cut the lower section of the seal from the bottom, and seal the hole in all three locations. Bend the seal in the desired position and install following normal sealing procedures. To make a downward turn with the seal, proceed as follows (figure 13-8, part c). Drill a M-inch hole along the seal at the location of the bend in the seal in a position 2/3H from the bottom of the seal. Cut a wedge from the underside of the seal by making two intersecting angled cuts (45 degrees) from the bottom of the seal to the ½-inch hole. Bend the seal down and install using normal sealing procedures.

c. *Expansion joint seals with an asphalt overlay.* The installation procedure is as follows (figure 13-9). Saw cut along lines parallel to and 1 foot on either side of the joint at a 60-degree angle away from the joint. Remove the asphalt inside the cut area and clean the concrete deck. Place a filler in the joint to maintain an area between the joint and asphalt. Apply an epoxy bonding compound to the cleaned concrete surface. Place latex modified concrete in the cut area and finish off flush with the existing asphalt. After the concrete has set, saw cut a joint to the required width and depth. Remove all debris and filler from the joint. Apply adhesive to the joint faces and install the seal.

d. *Expansion dam repair.*

(1) *Open armored joint.* This repair method can be used to spot repair a loose expansion dam or across the full length of armored, finger, and sliding joints. The repair involves removing a small portion of concrete, welding a steel strap to the dam and the deck's reinforcing steel, and replacing the concrete (figure 13-10, part a). The procedure is as follows. Identify the loose portions of the expansion joints requiring repair. Mark 12- by 8-inch rectangles immediately adjacent to the dam on 18-inch centers, and saw cut around the perimeter of the rectangle to a 1-inch depth. Remove the concrete inside the rectangle to a depth required to expose the deck's reinforcing steel with an air hammer. Cut a slot 1 inch wide into the dam and weld a ¼- by 1- by 12-inch steel Z-strap into the slot and to the deck's reinforcing steel. Sandblast the rectangular opening in the concrete and blow out all debris. Apply an epoxy bonding compound to the exposed concrete and fill the rectangle with a compacted latex modified portland cement concrete.

(2) *Sliding joint.* This repair can also be used on armored and finger joints. The repair places 1-inch diameter bolts on 2-foot centers through the bridge deck and the bolt is welded to the expansion dam. The bolt is anchored to the bottom of the deck with a steel plate, wedge washer, and nut (figure 13-10, part b). The repair procedure is as follows. Burn holes through the top flange of the expansion dam on 2-foot centers across the area to be repaired. Below the holes in the top flange, drill 1¼-inch hole through the deck at a 45-degree angle. Saw cut a 6- by 6-inch area, 1 inch deep around the exit of the drilled hole on the underside of the deck. Remove the concrete inside the scored area to the reinforcing steel. Place a 1-inch-diameter bolt through the hole in the flange and the deck. Cut the head of the bolt flush with the top flange and weld to the flange with a full penetration weld. The bolt should extend through

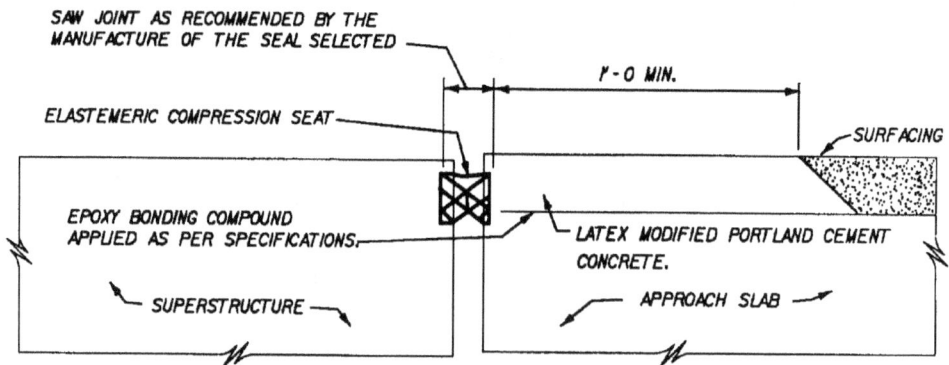

END DAM SYSTEM FOR
END BENT JOINT SEALS

Figure 13-9. Placing an elastomeric seal in an asphalt overlay.

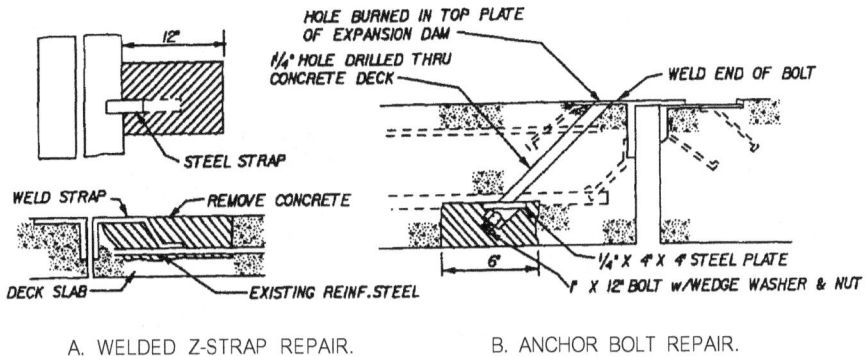

Figure 13-10. Expansion dam repairs.

the deck. Place a ½-inch-thick layer of latex modified concrete in the drilled hole around the bottom of the bolt. Position a ¼- by 4- by 4-inch steel plate with a 1¼-foot hole, add the wedge washer (45-degree wedge) and nut, then tighten. Apply an epoxy bonding compound to the 6- by 6-inch area and fill with latex modified concrete. Drill and tap the flange close to the bolt location and install a zerk fitting. Pump epoxy through the fitting to fill the voids under the expansion dam and around the bolt. Remove the zerk fittings and weld the hole closed. Grind the top flange smooth.

13-9. Abutments and wingwalls

The concrete in abutments and wingwalls may deteriorate from the effects of water, deicing chemicals, freeze cracking, or impact by debris which results in breaking off the edges or portions of the face. These conditions require that repairs be made to prevent continued deterioration, particularly increased spalling due to moisture reaching the rebar and causing corrosion. The following steps in the rehabilitation procedure are normally required (figures 13-11 and 13-12). Establish traffic control, as necessary. Excavate as required to set dowels and forms. Remove deteriorated concrete by chipping and blast cleaning. Drill and set tie screws and log studs to support the form work. Set reinforcing steel and forms. Apply epoxy-bonding agent to the concrete surface. Place concrete, cure, and remove forms. Install erosion control material as necessary.

13-10. Bridge seats

Problems often found in concrete bridge seats include deterioration of concrete, corrosion of the reinforcing bars, friction from the beam or bearing devices sliding directly on the seat, and the improper design of the seat which results in shear failure. During the preliminary planning stage,

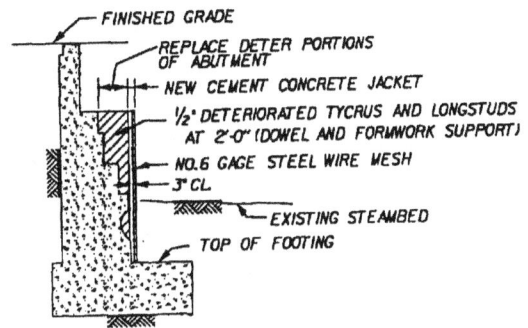

Figure 13-11. Repair of abutment and wingwall faces using a jacket.

the specific cause of the problems should be determined to properly repair the damage. A detailed plan of the jacking requirements should be made. Several repair procedures are as follows:

a. Abutment and cap seats (figure 13-13). Remove traffic from structure during jacking operations. Lift jacks in unison to prevent a concentration of stress in one area and possible damage to the superstructure. Restrict vehicle traffic during the repair as much as possible. Saw cut around concrete to be removed and avoid cutting reinforcing steel. Remove deteriorated concrete to the horizontal and vertical planes ensuring that sound concrete is exposed. Add reinforcing steel as required and construct forms to confine the new concrete. Apply bonding material to the prepared surface that will interface with the new concrete. Place and cure new concrete. Service, repair, or replace bearings as necessary.

b. Concrete cap extension. This repair restores adequate bearing for beams that deteriorated or sheared at the point of bearing by anchoring an extension to the existing cap. The procedure is as follows (figure 13-14). Locate and drill 6-inch-deep holes to form a grid in the existing cap and install

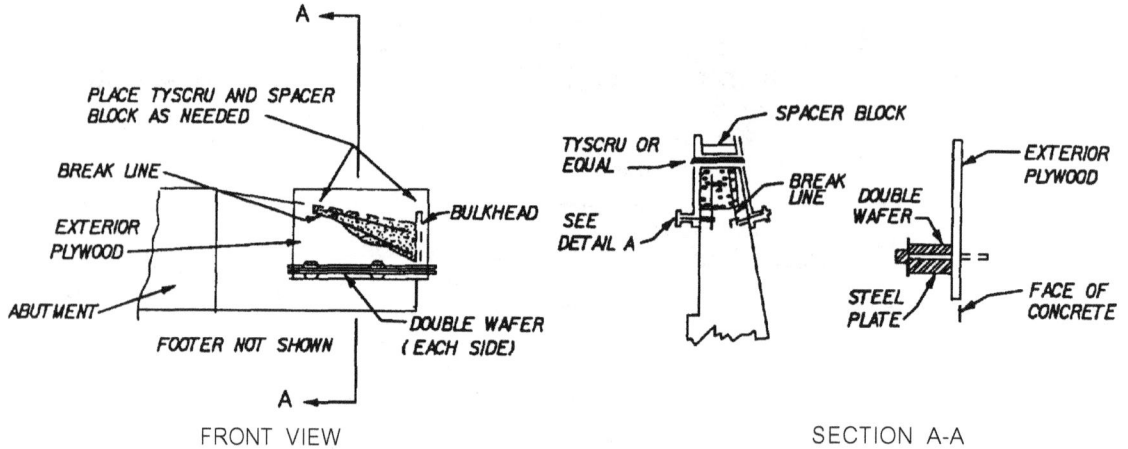

Figure 13-12. Repair of broken or deteriorated wingwalls.

concrete anchors that will accept a ¾-inch bolt. Place ¾- by 9-inch bolts 4 inches into the concrete anchors. Wire a reinforcing steel (No. 4 bars) grid to the inside head of the anchor bolts. Construct a form around the reinforcing steel grid with a minimum of 4 inches cover around the sides of the bolts and a minimum of 2 inches cover for the face of the extension. Place roofing paper against the bottom of the beam and place Class IV concrete in the form. Remove forms after 3 days. The extension should not carry any load during curing. Repair any damage to the end of the beam.

c. Beam saddle. The saddle restores bearing for beams and caps where they have deteriorated or been damaged in the bearing area. The procedure is as follows (figure 13-15). Obtain measurements of the cap and beam width, and have an engineer design the saddle. Prepare the top of the cap and the beams for good bearing contact with the saddle. Use neoprene bearing pads for contact points between the saddle and the beam and cap. Place saddle members in contact with the cap across the cap on each side of the beam. Drop one bolt through a hole at each end of the two saddle cap contact members, and bolt the saddle bearing members in place under the beam. Place neoprene bearing pads under the beam, and tighten the bolts in place. Install and tighten the remaining bolts.

13-11. Columns and piles

The typical repair involving concrete columns and piles is to place a concrete jacket around the member to protect it from further deterioration or to restore the structural integrity of the member. The repair can be made with a standard wood or metal form work which is removed after curing or a fiberglass form that remains in place and helps protect the surface of the member.

a. Standard formwork. This repair method is used for piles that have been damaged or deteriorated to the point that structural integrity of the member is in question. In this procedure, the pile is encased with a concrete jacket reinforced with epoxy coated reinforcing steel (figure 13-16). The construction procedure is as follows. Remove all deteriorated concrete to a sound base. Clean the pile to ensure proper bonding with the jacket. Sand-blast rebars to remove corrosion and rust. Place a rebar cage around the pile to reinforce the pile. Treat the inside face of the forms with a release agent, and set the forms for the concrete jacket around the rebar cage and pile. Dewater the forms and place Class III concrete. Allow a minimum of 72 hours for curing.

b. Fiberglass forms. These forms can be used to prevent further deterioration or restore structural integrity to piles and columns. The repair involves the encasement of the pile or column in the fiberglass form and filling the form with epoxy grout, cement grout, or Class III concrete. The fiberglass forms should be a minimum of 1/8-inch in thickness with noncorrosive standoffs and a compressible sealing strip at the bottom. Installation procedures will vary depending upon the degree and location of damage and the specific jacket manufacturer's recommendations. The basic

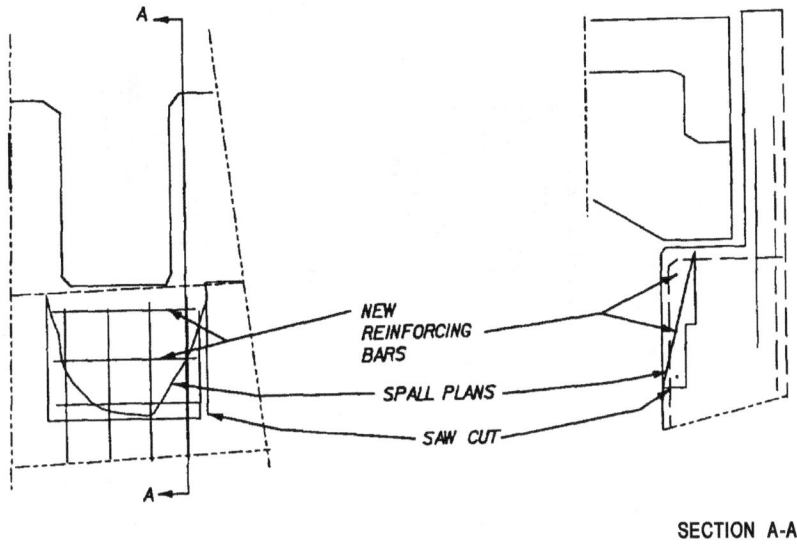

Figure 13-13. Typical repair of concrete bridge seats.

procedure for using fiberglass forms is as follows. Clean pile surfaces of materials which prevent bonding. Remove deteriorated concrete down to a sound concrete base. Sandblast exposed rebar to a "near-white metal" finish. Place the jacket form work around the pile. The form should have built in standoffs, concrete blocks attached to the inside face, or double-nut bolts placed through drilled holes to provide the proper standoff. Seal the form's joints and bottom with an epoxy compound. Place external bracing and bonding materials to help prevent form movement and bulging. De-water form and fill the space between the pile and form with the appropriate tiller. Cure, remove external bracing and banding, and clean any filler material from the form's faces.

13-12. Stringers and beams

Spall in beams can be dealt with in a similar manner as walls by constructing a form around the damaged area and replacing the lost concrete (paragraph 13-10a). Shotcrete is an excellent material to use in this type of repair. The major problem caused by concrete deterioration is the loss of effective reinforcing bar diameter due to corrosion. This is also true of cracks which allow water to penetrate to the reinforcing steel. Due to the criticality of reinforcing bars in beams, it is important to repair or replace any damaged reinforcing bars. There are three methods available to reestablish the reinforcing steel required for proper beam performance:

a. *Conventional repair.* Refer to paragraph 13-7.

b. *Conventional reinforcement.* Refer to paragraph 13-6a.

c. *Prestressing steel.* This method can be used to close a crack and/or provide external reinforcing steel to support the beam loading (figure 13-17). Install as follows. Clean the crack and any exposed rebar. Drill holes through the side of the beam (missing existing rebar) for the prestressing anchor at both ends of the beam. Install anchors on both sides and at both ends of the beam by running bolts through prepared holes in the anchors and beam. The anchor should be designed by an engineer and generally consist of a reinforced angle section with holes in the flanges to receive the through bolts and the tension tie. Connect the tension tie to each set of anchors on either side of the beam. Apply tension to the ties using a turnbuckle or torquing nuts. Tension should be applied across both sides of the beam evenly. Increase tension until the crack closes and seal with a flexible seal. This method can also be used in conjunction with a patch to replace deteriorated steel.

13-13. Decks

a. *Spall.* Spall repair on bridge decks can be broken into three categories based on the overall condition of the deck, as is shown further into this paragraph. The categories are identified according to the amount of spall, the extent of total deterioration (spall, delaminations, and corrosion poten-

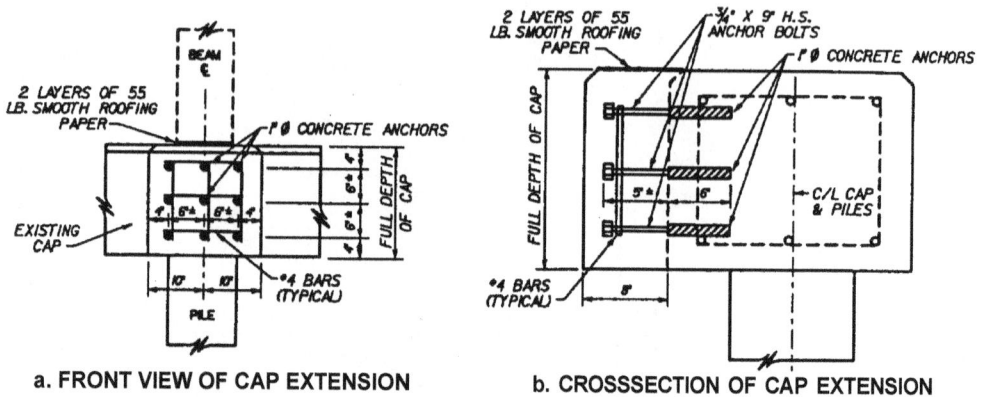

a. FRONT VIEW OF CAP EXTENSION

b. CROSSSECTION OF CAP EXTENSION

Figure 13-14. Concrete cap extension to increase bearing surfaces.

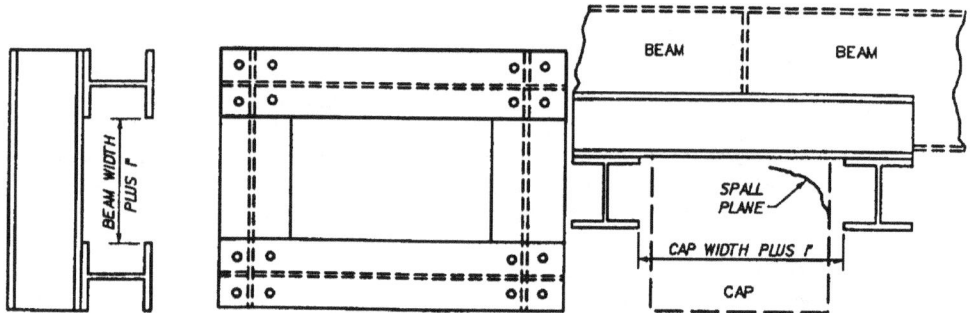

Figure 13-15. Typical beam saddle design using standard steel W-sections.

tials over   -0.35 volts), and total percentage of concrete samples containing at least 2 pounds of chloride per cubic yard of concrete:

(1) Extensive active corrosion.

(a) Spa11 covers more than 5 percent of total deck area.

(b) Deterioration covers more than 40 percent of total deck area.

(c) Chlorides high in over 40 percent of samples.

(2) Moderate active corrosion.

(a) Spa11 covers from 0 to 5 percent of total deck area.

(b) Deterioration covers from 5 to 40 percent of total deck area.

(c) Chlorides high in 5 to 40 percent of samples.

(3) Light to no active corrosion.

(a) No spa11 present.

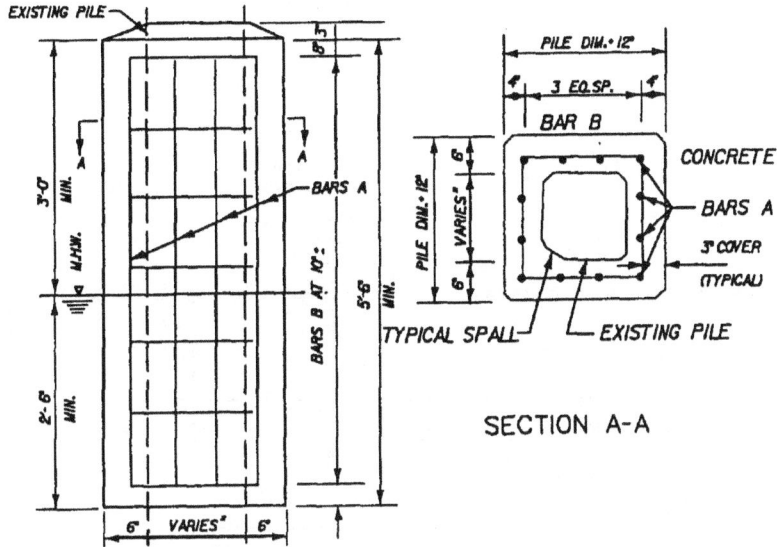

Figure 13-16. Standard concrete pile jacket with steel reinforcing cage.

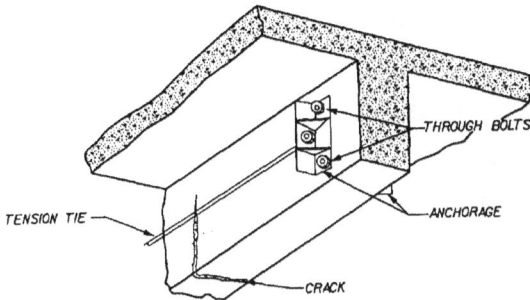

Figure 13-17. External prestressing strands used to close a crack.

(b) Deterioration covers from 0 to 5 percent of total deck area.

(c) Chlorides high in from 0 to 5 percent of samples.

In many cases the identifying characteristics of the moderate category will overlap with the other two categories, and a best judgment based on engineering, economics, and other factors must be used to establish the appropriate repair for the bridge. The deck repair procedures for each of these categories are outlined in table 13-2.

b. Cracks. Cracks in decks can be repaired as discussed in paragraph 13-6. In composite deck systems, prestressing techniques can be employed to help close cracks in the deck. The repair

Table 13-2. Bridge deck restoration procedures

Bridge Deck Restoration	Bridge Deck Deterioration					
	Light		Moderate		Extensive	
	Extend Life 10-15 years	Extend Life > 15 years	Extend Life 10-15 years	Extend Life > 15 years	Extend Life 10-15 years	Extend Life > 15 years
Remove and replace all areas of deterioration and chloride contaminated concrete.		X				
Remove all deteriorated concrete. Replace concrete in accordance with the selected protective system (see below).	X		X	X*	X	
Replace the entire deck.				X*		X
Protective Systems						
Membrane w/bc overlay.	X	X	X	X	X	X
Latex modified concrete overlay.	X	X	X	X	X	X
Cathodic protection.	X	X	X	X	X	X
Epoxy coated rebars.				X*		X

* Conduct a cost analysis to determine the most economical repair option.

consists of placing a tie rod across the deck crack. The tie rod is emplaced through drilled holes in the beams on either side of the cracked deck. The torquing nuts or turnbuckles on the tie rods are then tightened to close the crack in the deck (figure 13-18).

13-14. Replacement of concrete members

a. Decks. After the deteriorated portion of the deck has been removed, replacement decks can be cast in place or precast sections can be fixed into position. The advantage to using precast sections is that no form work is required, and smaller portions of the deck can be removed and replaced, thereby minimizing traffic disruption. The procedures for both types of deck replacement are outlined:

(1) *Cast in place.* Remove the deteriorated deck. Construct the form work for the new deck. Shore the form work to carry the dead load of the deck. Install reinforcing steel. Place the concrete and allow to cure long enough to achieve its design strength.

(2) *Precast.* Remove the deck area to be replaced with the precast section. Place the precast sections with a crane, being careful to overlap extended rebar when placing adjacent panels and wire together. Form around the connection between adjacent panels and apply an epoxy agent to interface between the sides of the panels and the new concrete.

b. Piles. Piles which cannot be jacketed can be replaced in much the same manner as timber piles (chapter 12).

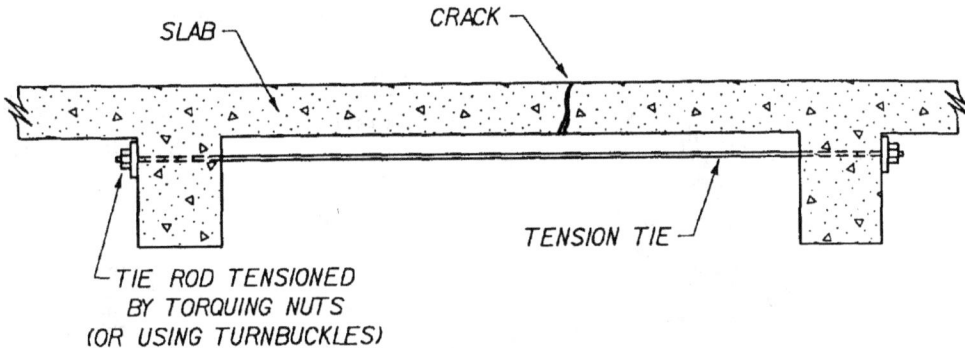

Figure 13-18. Closing a crack in a deck using prestressing steel.

Section III. UPGRADE CONCRETE BRIDGES

13-15. General upgrade methods

a. Substructure upgrade. The strength of the substructure can be increased by adding piles/columns to the pier or by placing a concrete jacket around the existing piles.

(1) *Jacketing (figure 13-19, part a).* The jacketing procedures for strengthening columns and piles are basically the same as the repair procedures discussed in section II of this chapter. For strengthening, the jacket must be extended the full length of the column and connected to the cap and footing. The jacket is connected to the existing cap and footing by grouting dowels into drilled holes.

(2) *Jacketing with spiraled reinforcement (figure 13-19, part b).* This method is used to improve the lateral strength of circular columns. It

involves wrapping tensioned prestressing wire around the existing column or applying a series of M-inch-diameter reinforcing bar hoops with turnbuckles at each end to pretension the hoops. A cover of shotcrete or cast-in-place concrete is then applied to the reinforcing steel.

(3) *Partial jacketing (figure 13-19, part c).* This technique involves attaching precast sections to the existing column using bolts or the casting of any additional reinforced concrete to the existing column that does not completely encase the column.

b. Span reduction. The common methods of shortening spans discussed in chapter 10 also apply for concrete bridges. These methods include intermediate piers, A-frames, or knee braces.

c. Posttensioning. All the posttensioning techniques shown in table 10-3 can be applied to concrete bridges. The connections and components for these systems should be designed by a structural engineer.

d. Add stringers. Additional concrete stringers can be cast in place under the deck and connected to the deck with dowels. The added dead load may reduce the benefit of the additional stringers and may prove uneconomical to construct. Another alternative is the addition of steel stringers between each of the existing concrete stringers. The abutments and caps can have seats constructed on them for the steel members.

13-16. Strengthen individual members

Individual concrete members can be strengthened by jacketing, inserting reinforcing bars, prestressing, or by external reinforcement. Members are generally strengthened to counter increases in shear or flexural forces.

a. Shear reinforcement of beams.

(1) *External reinforcement (figure 13-20).* Install external reinforcement as follows. Run channel sections across the top of the beam or attach channel sections to either side of the upper portion of the beam using through bolts in drilled holes. Run channel sections across the bottom of the beam and place tie rods through prepared holes in the top and bottom channels. Use torque nuts on either end of the rods to prestress the rods. The prestressed rods installed at specified spacings provide an increase in the shear capacity of the beam.

(2) *Reinforcing bar insertion (figure 13-2).* Insert bars as follows. Seal all cracks with silicone rubber. Mark the girder centerline on the deck. Locate the transverse deck reinforcement. Drill 45-degree holes avoiding the reinforcing bars. Pump epoxy into the holes and cracks, and insert the reinforcing bars into the epoxy-filled holes.

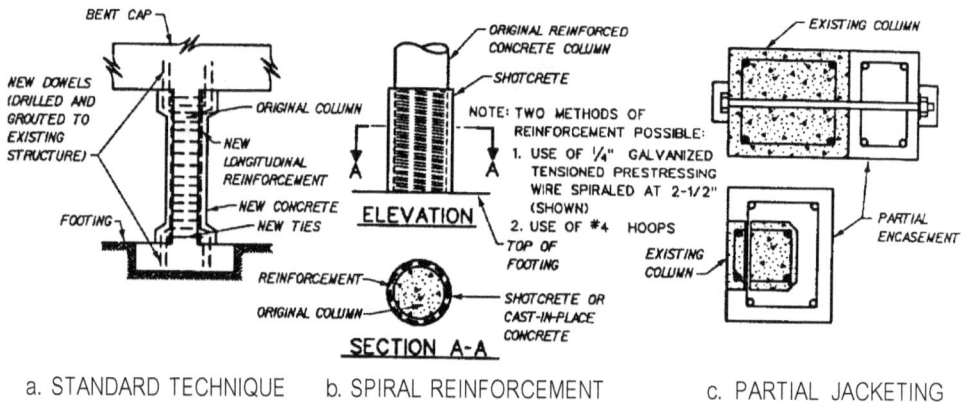

a. STANDARD TECHNIQUE   b. SPIRAL REINFORCEMENT   c. PARTIAL JACKETING

Figure 13-19. Jacketing of concrete columns.

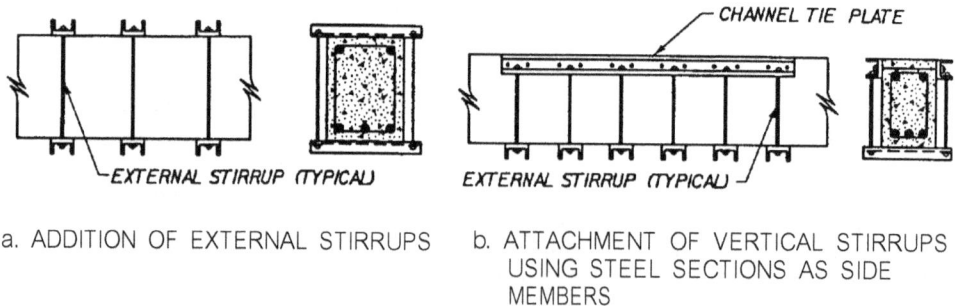

a. ADDITION OF EXTERNAL STIRRUPS   b. ATTACHMENT OF VERTICAL STIRRUPS USING STEEL SECTIONS AS SIDE MEMBERS

Figure 13-20. External shear reinforcement for concrete beams.

b. Shear reinforcement of box beams.

(1) *External reinforcement (figure 13-21).* Install as follows. Drill holes for the tendons through or just outside the web of the box beam. Counter sink the drill holes. Install the prestressing tendons. Inject epoxy into any existing cracks. Tighten the nuts on the tendon to prestress the beam, carefully following design specifications to avoid overstressing the beam. Inject epoxy around the tendons, and dry pack the countersink holes flush with the concrete surface.

(2) *Web reinforcement (figure 13-22).* Install as follows. Locate and drill holes on the inside face of the web of the box-beam. Clean the holes of dust and debris, and set concrete anchors in place. Place anchor bolts with a spacer and steel plate attached. Weld rebar to the steel plate on either side of the bolt. Attach horizontal reinforcing bars to the vertical rebar.

c. Flexural reinforcement of beams. The two most common methods of strengthening concrete beams in flexure are adding steel cover plates to the beam's tension face and partial encasement of the beam in reinforced concrete.

(1) *Steel channel cover plate (figure 13-23).* Install as follows. Remove dirt and any foreign material, sand blast, and then remove any debris with compressed air. Locate the beam's stirrups and longitudinal steel. Mark the location of the drill holes to miss the reinforcing steel and still be above the bottom row of longitudinal steel. Drill holes through the concrete beam. Drill or cut holes in the steel channel to match the holes in the beam. Position the channel into place and install bolts to hold it in place. Inject epoxy resign into the remaining holes and install the bolts. Take out the positioning bolts, inject epoxy, and reinstall. Seal between the channel and the concrete beam.

(2) *Partial jacketing.* The purpose of the jacketing in this case is to provide cover for the additional reinforcing bars which are positioned to help carry the flexural load. The longitudinal

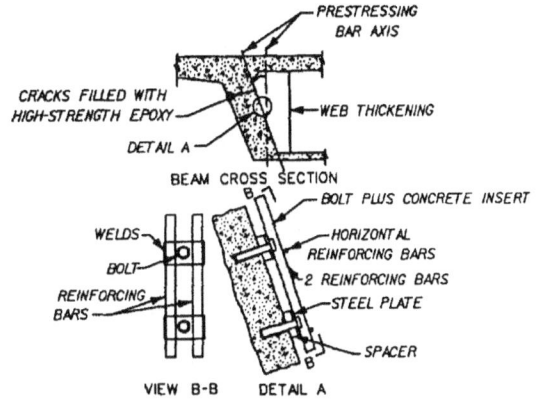

Figure 13-22. Web reinforcement of boxbeams.

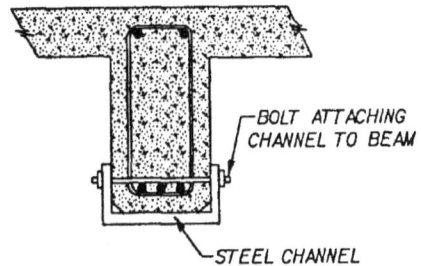

Figure 13-23. Steel channel used to reinforce beams

reinforcing bars are held in place by the vertical bars that are attached to the existing beam (figure 13-24, part a), placed through the beam (figure 13-24, part b), or run through the existing deck (figure 13-25, part c). The concrete cover can be cast in place or shotcreted.

(3) *External plates.*

(a) This method involves bonding mild steel plates to the concrete member. This composite system relies on the effectiveness of the bonds between the concrete/adhesive and the adhesive/steel surfaces. This technique should not be used on members with reinforcement corrosion or high concentrations of chloride ions.

(b) Design considerations to avoid brittle cracking of the concrete cover are as follows: use steel plates with a minimum width/thickness ratio of 40 to reduce horizontal shears; limit the nominal elastic horizontal shear stress in the adhesive to the tensile capacity of the concrete for an applied safety factor.

(c) This strengthening method should be conducted by trained personnel using the following procedure. Clean and sandblast the steel plates and apply an epoxy primer paint adhesive. Sand

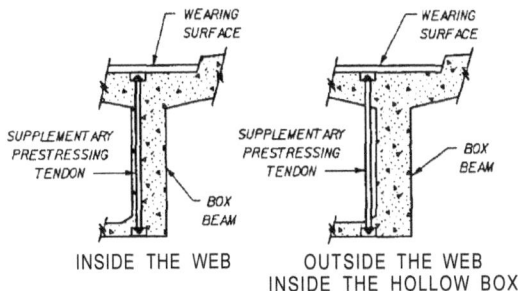

Figure 13-21. Reinforcing boxbeams for shear.

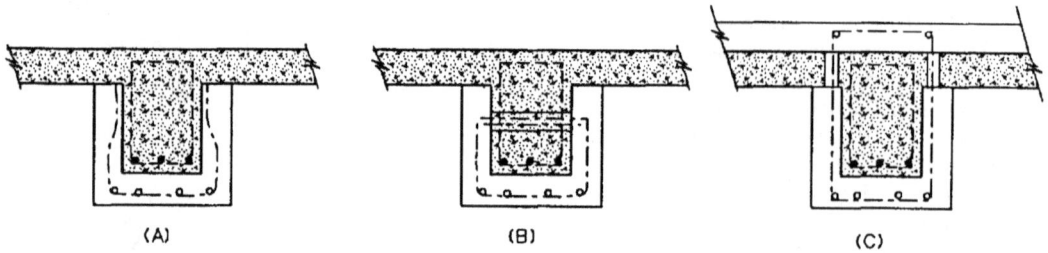

(A)   (B)   (C)

Figure 13-24. Concrete beams reinforced with concrete sleeves.

blast the concrete to remove the surface laitance and to expose coarse aggregates. Drill holes in the slab or beam to emplace anchor bolts for the ends of the steel plates. Mix adhesive and apply to the steel plate with a thickness between 1/16th to 1/8th inch. The adhesive should be thicker along the centerline of the plate to prevent air pockets from forming between the plate and concrete faces. Place the steel plate on the underside of the beam or slab and bolt the ends to hold into place. Supplemental bolting is used at the plate ends to reduce peel failures and hold the plate in place in case of failure. Apply pressure along the full length of the plate using secondary supports and wedges. Tighten the end bolts into place and allow the adhesive to cure (normally 24 hours). Remove secondary supports and paint the steel plate.

13-17. Prestressed concrete members

By design, prestressed concrete members behave differently than conventionally reinforced members. As a result, many of the methods discussed in the preceding paragraphs do not apply to prestressed concrete members. Due to its complexity, the repair of prestressed concrete members is not covered in this manual. The following reports by the Transportation Research Board should be referred to for the evaluation and repair of prestressed concrete members:

(1) National Cooperative Highway Research Program (NCHRP) Report Number 226, "Damage Evaluation and Repair Methods for Prestressed Concrete Bridge Members."

(2) NCHRP Report Number 280, "Guidelines for Evaluation and Repair of Prestressed Concrete Bridge Members."

APPENDIX A

REFERENCES AND BIBLIOGRAPHY

A-1. References

Government Publications

Departments of the Army, Navy, and Air Force

AR 420-72	Surfaced Areas, Bridges, Railroad Track and Associated Appurtenances
FM 5-446	Military Nonstandard Fixed Bridging
TM 5-628	Railroad Track Standards
FM 5-277	Bailey Bridge

Coast Guard

Pamphlet CG204	Aids to Navigation

Department of Transportation, Federal Highway Administration

Bridge Inspector's Training Manual 70
Inspection of Fracture Critical Bridge Members

Nongovernment Publications

American Association of State Highway Transportation Officials (AASHTO), 444 North Capitol Street, Washington, DC 20001

Standard Specifications for Highway Bridges, Fourteenth Edition, 1989
Manual on Uniform Traffic Control Devices
Manual for Maintenance Inspection of Bridges

National Cooperative Highway Research Program (NCHRP), Transportation Research Board, National Research Council, Washington, DC

Report 280	Guidelines for Evaluation and Repair of Prestressed Concrete Bridge Members, December 1985
Report 293	Damage Evaluation and Repair Methods for Prestressed Concrete Bridge Members, November 1980

A-2. Bibliography

Manuals

AASHTO Maintenance Manual, 1987. American Association of State Highway and Transportation Officials, Washington, DC (1987).

AASHTO Manual for Bridge Maintenance, 1987. American Association of State Highway and Transportation Officials, Washington, DC (1987).

American Concrete Institute (ACI) Manual of Concrete Practice, 1988. American Concrete Institute, Detroit, MI.

Annual Book of American Society for Testing and Materials (ASTM) Specification, 1990. American Society for Testing and Materials, 1961 Race Street, Philadelphia, PA.

Guide Specifications for Strength Evaluation of Existing Steel and Concrete Bridges, 1989. American Association of State Highway and Transportation Officials, Washington, DC (1989).

Guide Specifications for Fracture Critical Non-Redundant Steel Bridge Members, 1978. American Association of State Highway and Transportation Officials, Washington, DC (1989).

Manual for Railway Engineering, American Railway Engineering Association (AREA), Washington, DC (1989).

Manual for Bridge Maintenance Planning and Repair Methods, Volume I. Florida Department of Transportation, State Maintenance Office.

Technical Publications-Departments of the Army, Navy, and Air Force

FM 5-742   Concrete and Masonry
FM 5-551   Carpentry
TM 5-744   Structural Steel Work
TM 5-622/MO-104/AFM 91-34   Maintenance of Waterfront Facilities

Reports

National Cooperative Highway Research Program, Report 293.   "Methods of Strengthening Existing Highway Bridges."   Transportation Research Board, National Research Council, Washington, DC, September 1987.

National Cooperative Highway Research Program, Report 333.   "Guidelines for Evaluating Corrosion Effects in Existing Steel Bridges."   Transportation Research Board, National Research Council, Washington, DC, December 1990.

Journal Articles

American Concrete Institute Committee 201, "Guide to Durable Concrete,"   *Journal of the American Concrete Institute,* Number 12, Proceedings Volume 74, pages 573-609, December 1977.

APPENDIX B

SUGGESTED ITEMS FOR ARMY ANNUAL AND AIR FORCE BIANNUAL BRIDGE INSPECTIONS

BRIDGE INSPECTION ITEMS

Include the following items:
1. Installation.
2. Bridge number.
3. Location.
4. Date inspected.
5. Existing bridge classification (if applicable).

For the following components, address each appropriate inspection item and make notes of any observed deficiencies and recommendations:

A. Timber Abutments
1. Signs of settlement.
2. Rusting of steel rods.
3. Decay of end dam, wingpost, post, and/or cap.
4. Deterioration of block (bearing and anchor).
5. Decay of sill and footing.
6. Loose timbers.
7. Decay of breakage of piles (wing or bearing).

B. Steel Pile Abutments
1. Settlement.
2. Rusting of end dam, pile and/or cap.
3. Section loss of steel members.
4. Missing, loose, or rusting bolts.

C. Concrete Abutments, Wingwalls, and Retaining Walls
1. Settlement.
2. Proper function of weep holes.
3. Cracking or spalling of bearing seats.
4. Deterioration of cracking of concrete.
5. Exposed reinforcing steel.

D. Timber Piers and Bents
1. Settlement.
2. Decay of caps, bracing, scabbing, or corbels.
3. Missing posts or piles.
4. Decay of posts or piles.
5. Debris around or against piers.
6. Section loss of sills or footings.
7. Erosion around piers.
8. Rusting of wire-rope cross bracing.
9. Loose or missing bolts.
10. Splitting or crushing of the timber when:
    a. The cap bears directly upon the cap, or
    b. Beam bears directly upon the cap.
11. Excessive deflection or movement of members.

E. Steel Piers and Bents
1. Settlement or misalignment.
2. Rusting of steel members or bearings.

3. Debris.

4. Rotation of steel cap due to eccentric connection.

5. Braces with broken connections or loose rivets or bolts.

6. Member damage from collision.

7. Need for painting.

8. Signs of excessive deflection or movement of members.

F. Concrete Piers and Bents

1. Settlement.

2. Deterioration or spalling of concrete.

3. Cracking of pier columns and/or pier caps.

4. Cracking or spalling of bearing seats.

5. Exposed reinforcing steel.

6. Debris around piers or bents.

7. Section loss of footings.

8. Erosion around piers.

9. Collision damage.

G. Concrete (girders, beams, frames, etc.)

1. Spalling (give special attention to points of bearing).

2. Diagonal cracking, especially near supports.

3. Vertical cracks or disintegration of concrete, especially in the area of the tension steel.

4. Excessive vibration or deflection during vehicle passage.

5. Corrosion or exposure of reinforcing steel.

6. Corroded, misaligned, frozen, or loose metal bearings.

7. Tearing, splitting, bulging of elastomeric bearing pads.

H. Timber (trusses, beams, stringers, etc.)

1. Broken, deteriorated, or loose shear connectors.

2. Failure, bowing, or joint separation of individual members of trusses.

3. Loose, broken, or worn planks on the timber deck.

4. Improper functioning of members.

5. Rotting or deterioration of members.

I. Steel (girders, stringers, floor beams, diaphragms, cross frames, portals, sway frames, lateral bracing, truss members, bearing and anchorage, eyebars, cables, and fittings)

1. Corrosion and deterioration along:

 a. Web flange.

 b. Around bolts and rivets heads.

 c. Under deck joints.

 d. Any other points which may be exposed to roadway drainage.

 e. Eyebars, cables, and fittings.

2. Signs of misalignment or distortion due to overstress, collision, or fire.

3. Wrinkles, waves, cracks, or damage in the web and flange of steel beam, particularly near points of bearing.

4. Unusual vibration or excessive deflection occurring during the passage of heavy loads.

5. Frozen or loose bearings.

6. Splitting, tearing, or bulging in elastomeric bearing pads.

J. Concrete Appurtenances

1. Cracking, scaling, and spalling on the:

 a. Deck surface.

 b. Deck underside.

 c. Wearing surface (map cracking, potholes, etc.).

NOTE: If deterioration is suspected, remove a small section of the wearing surface in order to check the condition of the concrete deck.

2. Exposed and/or rusting reinforcing steel.

3. Loose or deteriorated joint sealant.

4. Adequacy of sidewalk drainage.

5. Effect of additional wearing surfaces on adequacy of curb height.

K. Timber Appurtenances

1. Loose, broken, or worn planks.

2. Evidence of decay, particularly at the contact point with the stringer where moisture accumulates.

3. Excessive deflection or loose members with the passing of traffic.

4. Effect of additional wearing surfaces on adequacy of curb height.

L. Steel Appurtenances (including but not limited to decks, gratings, curbs, and sidewalks)

1. Corroded or cracked welds.

2. Slipperiness when deck or steel sidewalk is wet.

3. Loose fasteners or loose connections.

4. Horizontal and vertical misalignment and/or collision damage.

M. Masonry Bridges

1. Settlement.

2. Proper function of weep holes.

3. Collision damage.

4. Spalling or splitting of rocks.

5. Loose or cracked mortar.

6. Plant growth, such as lichens and ivy, attaching to stone surfaces.

7. Marine borers attacking the rock and mortar.

N. Miscellaneous

1. Existence and appropriateness of bridge classification signs.

2. Condition of approachments.

3. Leaks, breaks, cracks, or deterioration of pipes, ducts, or other utilities.

4. Damaged or loose utility supports.

5. Wear or deterioration in the shielding and insulation of power cables.

APPENDIX C

SUGGESTED ITEMS FOR
ARMY TRIENNIAL AND EVERY THIRD AIR FORCE BIANNUAL
BRIDGE INSPECTIONS

BRIDGE INSPECTION ITEMS

Include the following items:

A. General Information to Include
1. Bridge name.
2. Location.
3. Date of inspection.
4. Design load (if known).
5. Military load classification (if known).
6. Date built.
7. Traffic lanes.
8. Transverse section (describe or sketch).
9. Structure length.
10. No. of spans.
11. Plans available.
12. Inspection records.
   a. Year inspected.
   b. Inspector.
   c. Qualification.
13. Bridge description.
   a. Floor system.
   b. Beams.
   c. Girders.
   d. Stringers.
   e. Trusses.
   f. Suspension.
   g. Piers.
   h. Abutment A.
   i. Abutment B.
   j. Foundation.
   k. Piers or bents.
      (1) Caps.
      (2) Posts or columns.
      (3) Footings.
      (4) Piles.
      (5) Other.
   l. Deck:
      (1) Wearing surface.
      (2) Curb.
      (3) Railings.
      (4) Sidewalk.
      (5) Other.

B. Bridge Components Rating Information
The following items may be rated using the suggested ratings from part C of this appendix. Descriptive remarks may also be included.
   1. Traffic safety features.
   a. Bridge railing.

   b. Transitions.

   c. Approach guardrail.

   d. Approach guardrail terminal.

2. Deck.

   a. Wearing surface.

   b. Deck structural condition.

   c. Curbs.

   d. Median.

   e. Sidewalk.

   f. Parapet.

   g. Railings.

   h. Drains.

   i. Lighting.

   j. Utilities.

   k. Expansion joints.

3. Load bearing components.

   a. Bearing devices.

   b. Stringers.

   c. Girders or beams.

      (1) General.

      (2) Cross frames.

      (3) Bracing.

   d. Floor beams.

   e. Trusses.

      (1) General.

      (2) Portals.

      (3) Bracing.

   f. Paint.

4. Abutments.

   a. Wings.

   b. Backwall.

   c. Bearing seats.

   d. Breast wall.

   e. Weep holes.

   f. Footing.

   g. Piles.

   h. Bracing.

   i. Erosion or scour.

   j. Settlement.

5. Piers/bents or pile bents.

   a. Caps.

   b. Bearing seats.

   c. Column, stem, or wall.

   d. Footing.

   e. Piles.

   f. Bracing.

   g. Erosion or scour.

   h. Settlement.

6. Channel and channel protection.

   a. Channel scour.

   b. Embankment erosion.

   c. Drift.

   d. Vegetation.

   e. Fender system.

   f. Spur dikes and jetties.

g. Rip rap.

h. Adequacy of opening.

7. Approach.

    a. Alignment.

    b. Approach.

    c. Relief joints.

    d. Approach.

      (1) Guardrail.

      (2) Pavement.

      (3) Embankment.

C. Suggested Component Ratings

1. Traffic Safety Features.

Code	Description
0	Inspected feature DOES NOT currently meet acceptable standards or a safety feature is required and NONE IS PROVIDED.
1	Inspected feature MEETS currently acceptable standards.
N	NOT APPLICABLE

2. Superstructure, Substructure, Channel and Channel Protection, and Approach.

Code	Description
N	NOT APPLICABLE
9	EXCELLENT CONDITION
8	VERY GOOD CONDITION-no problems noted.
7	GOOD CONDITION-some minor problems.
6	SATISFACTORY CONDITION-structural elements show some minor deterioration.
5	FAIR CONDITION-all primary structural elements are sound but may have minor section loss, cracking, spalling or scour.
4	POOR CONDITION-advanced section loss, deterioration, spalling or scour.
3	SERIOUS CONDITION-loss of section, deterioration, spalling or scour have seriously affected primary structural components. Local failures are possible. Fatigue cracks in steel or shear cracks in concrete may be present.
2	CRITICAL CONDITION-advanced deterioration of primary structural elements. Fatigue cracks in steel or shear cracks in concrete may be present or scour may have removed substructure support. Unless closely monitored it may be necessary to close the bridge until corrective action is taken.
1	"IMMINENT" FAILURE CONDITION-major deterioration or section loss present in critical structural components or obvious vertical or horizontal movement affecting structure stability. Bridge is closed to traffic but corrective action may put back in light service.
0	FAILED CONDITION-out of service-beyond corrective action.

3. Supplemental for Channel and Channel Protection (Use in conjunction with part 2 above).

Code	Description
N	NOT APPLICABLE bridge is not over a waterway.
9	There are no noticeable or noteworthy deficiencies which affect the condition of the channel.
8	Banks are protected or well vegetated. River control devices such as spur dikes and embankment protection are not required or are in a stable condition.
7	Bank protection is in need of minor repairs. River control devices and embankment protection have little minor damage. Banks and/or channel have minor amounts of drift.
6	Bank is beginning to slump. River control devices and embankment protection have widespread minor damage. There is minor stream bed movement evident. Debris is restricting the waterway slightly.
5	Bank protection is being eroded. River control devices or embankment have major damage. Trees and brush restrict the channel.

Code	Description
4	Bank and embankment protection is severely undermined. River control devices have severe damage. Large deposits of debris are in the waterways.
3	Bank protection has failed. River control devices have been destroyed. Stream bed aggradation, degradation, or lateral movement has changed the waterway to now threaten the bridge or approach roadway.
2	The waterway has changed to the extent the bridge is near a state of collapse.
1	Bridge is closed because of channel failure. Corrective action may put it back in light service.
0	Bridge is closed because of channel failure. Replacement is necessary.

4. Supplemental for Approach Roadway Alignment (Use in conjunction with part 2 above):

Code	Description
8	Speed reduction is NOT required.
6	A VERY MINOR speed reduction is required.
3	A SUBSTANTIAL speed reduction is required.

www.ingramcontent.com/pod-product-compliance
Lightning Source LLC
Chambersburg PA
CBHW051214200326
41519CB00025B/7115